Advances in Technological Innovations in Higher Education

The evolution of technology in education can no longer be comprehended simply by looking at the use of computers and networks. Technology is not just a supplementary tool to the conventional method of education. Education has to undergo a complete transformation with technological innovations for the sustainability of quality education as a system and not in silos. The sustainability in education also necessitates a more workable strategy to realize socially viable educational policies and practices which can focus more on personalized learning. Due to various factors like emerging technologies; changing needs of the learners; policy reforms for enhancing employability; and emphasis on uninterrupted education as in the case of the pandemic scenario of COVID-19, there is a need to steer a major transition in the education system. The education system has to be real and proficient for it to be instrumental to nurture an informed and knowledgeable society. This book on technological innovations in higher education is organized, largely, based on the diversity of higher education ecosystems that are supported by technological innovations. Various author viewpoints give insights into the full potential of technology as well as its risks in interrelated areas of higher education to work towards sustainability of value-based quality education across the globe.

Innovations in Intelligent Internet of Everything (IoE)
Series Editor: Fadi Al-Turjman

Computational Intelligence in Healthcare: Applications, Challenges, and Management
Meenu Gupta, Shakeel Ahmed, Rakesh Kumar, and Chadi Altrjman

Blockchain, IOT and AI technologies for Supply Chain Management
Priyanka Chawla, Adarsh Kumar, Anand Nayyar, and Mohd Naved

Renewable Energy and AI for Sustainable Development
Editors: Sailesh Iyer, Anand Nayyar, Mohd Naved, and Fadi Al-Turjman

Advances in Technological Innovations in Higher Education: Theory and Practices
Editors: Adarsh Garg, B V Babu, and Valentina E Balas

For more information about the series, please visit: https://www.routledge.com/Innovations-in-Intelligent-Internet-of-Everything-IoE/book-series/IOE

Advances in Technological Innovations in Higher Education
Theory and Practices

Edited by
Adarsh Garg
B V Babu
Valentina E Balas

Taylor & Francis Group
Boca Raton London New York

CRC Press is an imprint of the
Taylor & Francis Group, an **informa** business

Designed cover image: © Shutterstock

First edition published 2024
by CRC Press
2385 NW Executive Center Drive, Suite 320, Boca Raton FL 33431

and by CRC Press
4 Park Square, Milton Park, Abingdon, Oxon, OX14 4RN

CRC Press is an imprint of Taylor & Francis Group, LLC

© 2024 selection and editorial matter, Adarsh Garg, B V Babu, and Valentina E Balas; individual chapters, the contributors

Reasonable efforts have been made to publish reliable data and information, but the author and publisher cannot assume responsibility for the validity of all materials or the consequences of their use. The authors and publishers have attempted to trace the copyright holders of all material reproduced in this publication and apologize to copyright holders if permission to publish in this form has not been obtained. If any copyright material has not been acknowledged please write and let us know so we may rectify in any future reprint.

Except as permitted under U.S. Copyright Law, no part of this book may be reprinted, reproduced, transmitted, or utilized in any form by any electronic, mechanical, or other means, now known or hereafter invented, including photocopying, microfilming, and recording, or in any information storage or retrieval system, without written permission from the publishers.

For permission to photocopy or use material electronically from this work, access www.copyright.com or contact the Copyright Clearance Center, Inc. (CCC), 222 Rosewood Drive, Danvers, MA 01923, 978-750-8400. For works that are not available on CCC please contact mpkbookspermissions@tandf.co.uk

Trademark notice: Product or corporate names may be trademarks or registered trademarks and are used only for identification and explanation without intent to infringe.

ISBN: 978-1-032-45380-4 (hbk)
ISBN: 978-1-032-45381-1 (pbk)
ISBN: 978-1-003-37669-9 (ebk)

DOI: 10.1201/9781003376699

Typeset in Sabon
by MPS Limited, Dehradun

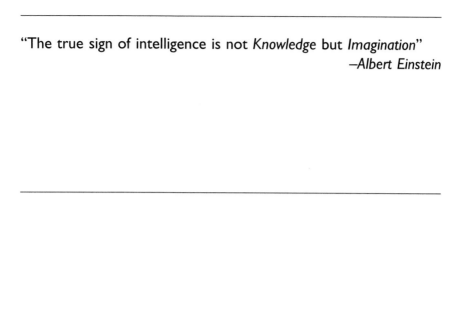

Contents

Preface	ix
Editor Biographies	xii
Contributors	xiv
Acknowledgements	xvii
Abbreviations	xviii

1 Rehumanizing education in the age of technology   1
SUBHAJIT GHOSH

2 Transitioning pedagogies in evolving India: Critical analysis of skills, knowledge, and wisdom with respect to implementation of NEP 2023   16
NEERJA ASWALE, NIRANJAN KULKARNI, RASHMIL SINGH, AND ARCHANA SINGH

3 Content and usability of MOOC platforms for e-learning: An evaluation in higher education   28
ADARSH GARG, P PRADEEP KUMAR, AND RAVINDER RENA

4 Machine learning in medical imaging: A comprehensive study   42
DEBANJANA GHOSH AND SRILEKHA MUKHERJEE

5 Platform with anonymity for students to foster in-class participation   51
ABHISHEK DEUPA AND RUQAIYA KHANAM

6 Speech emotion analyzer using deep learning	59
NAAZNEEN AHMED, RITIKA CHAMARIA, DIPTARKA PAUL, SUBHAM SARANGI, ISHIKA AGARWAL, YAMINI SHARMA, AND SRILEKHA MUKHERJEE	
7 Technology-enhanced personalized learning in higher education	71
RAVI KANT VERMA, SATYENDRA GUPTA, AND SVITLANA ILLINICH	
8 Environment for personalized learning	93
JAGJIT SINGH DHATTERWAL, KULDEEP SINGH KASWAN, AND SUNIL KUMAR BHARTI	
9 AI in personalized learning	103
KULDEEP SINGH KASWAN, JAGJIT SINGH DHATTERWAL, AND RUDRA PRATAP OJHA	
10 Transformative innovation in education	118
SAPAN ADHIKARI	
11 eSCOOL: A virtual learning platform	139
RUQAIYA KHANAM, SHRAIYASH PANDEY, SHRISHTI CHOUDHARY, AND ABHIK KUMAR DE	
12 Post pandemic technology assisted teaching and learning: A perspective on self-directed learning	151
SHREYA VIRANI AND SARIKA SHARMA	
13 Education 5.0: An overview	168
B V BABU	

Preface

It is believed that higher education systems should comply with innovation patterns and innovate themselves. Accordingly, higher education should change the model of working so that learners can create suitable knowledge which can be used in future products/services. So, technological innovations in education can be revealed with a more holistic approach by developing a standard ecosystem of teaching-learning. This ecosystem will revolve around the theory, pedagogy, content, context and capability as a system. Innovations in education with emerging technologies can raise productivity and efficiency of learning so as to cater to the needs of every individual learner, exclusively. Further, the strategic implementation of the ecosystem will bring in significant and positive change in teaching-learning in higher education in an international context.

The technological innovations so far have not been able to sustain the learner's predisposition to successfully complete the learning. The technology evolution in education can no longer be apprehended simply by looking at the use of computers and networks. Most of the research talks about technology as a supplementary tool to the conventional method of education. An extensive comparison has been made between past and current practices. But what is vital here is to undergo a complete transformation with technological innovations for sustainability of quality education as a system and not in silos. Sustainability in education also necessitates a more workable strategy to realize socially viable educational policies and practices which can focus more on personalized learning.

Due to various factors like emerging technologies; changing needs of the learners; policy reforms for enhancing employability; and emphasis on uninterrupted education as in the case of current pandemic scenario of COVID-19, there is a need to steer a major transition in education system. The education system has to be real and proficient for it to be instrumental to nurture the knowledge-society.

This book on advances in technological innovations in higher education is organized, largely, on the diversity of higher education ecosystem that are supported by technological innovations. Various viewpoints of authors give insights into the full potential of technology as well as its risks in interrelated areas of higher education to work towards the sustainability of value-based quality education, globally. The insights are based on qualitative/quantitative empirical research and the theoretical analysis as follows:

Chapter 1 throws light on changing expectations from education. Especially, expectations from Higher Education sectors include not limiting only to Teaching, doing Research, or serving the community, but also innovating and bringing out transformation in the industrial processes. The chapter provides insights into how minimizing interactions between humans and leaving decision-making capability solely to technology can bring about undesirable consequences.

Chapter 2 witnesses a paradigm shift in teaching pedagogies showcasing the ups and downs of the Indian Education System. It focuses on the timelines of the education system with respect to challenges faced in each era in terms of Skills Knowledge and Wisdom and implementation of NEP in 2023 in India. The critical analysis of the transitioning Pedagogies in the education system is well articulated and presented through a qualitative research methodology.

Chapter 3 emphasizes technological advances in higher education which is gradually moving from classroom teaching to a more advantageous online platform. It explains the Massive Open Online Course platforms, which define a site for efficient communication, collaboration, creativity and critical thinking. It provides a critical investigation and review of the existing MOOC platforms regarding its content and usability.

Chapter 4 discusses the use of Machine Learning as a boon to generate automated or semi-automated models, to analyse the data better and faster with acceptable accuracy in higher education in medicine. It shows a huge need to design efficient and reliable semi-automated or automated models for the analysis of medical images to get a machine interpretation in every day medical practice.

Chapter 5 talks about the importance of participation of students in a classroom in the teaching and learning process. It helps them better understand the course material and is good feedback to teachers. It proposes a platform which allows students to remain anonymous among others while being able to participate in active interaction with teachers and giving them feedback. This would increase the propensity of students to participate in classroom discussion and also reveal their own responses in the classroom.

Chapter 6 describes an "Emotion Speech Analyzer" to build a machine leaning model which will be able to detect human emotions whenever needed. This composition describes the creation of a machine learning model for performing real time analysis of speech for detecting emotional states of human beings and accessing their quality of speech.

Chapter 7 addresses the concept of personalized learning which is the approach of matching educational material that connects a learner's prior understanding with new information with their prior knowledge, experiences, and skills. Learning Record Store (LRS) and other technological solutions like experience-tailored learning make it possible to digitize the learners' experiences. Using a mix of advanced search and tailored engines, the information may be used to create a personalized learning experience for future learning activities.

Chapter 8 portrays the environment for personalized learning for science students. It shows students' need for science materials to supplement their current course load. It considers how students are utilizing them, how that may vary from the recommended practice indicated in the literature, and what that may mean for the use of technological devices in the classroom.

Chapter 9 is an endeavour to show the use of AI in personalized learning. It explores various AI-powered tools and techniques used in personalized learning environments, such as intelligent tutoring systems, recommendation engines, and adaptive assessments. AI in personalized learning facilitates adaptive content delivery, enabling students to learn at their own pace and according to their individual strengths and weaknesses. It also portrays the need to address the issues of life privacy concerns, ethical considerations, and potential biases in AI algorithms, to ensure the responsible and equitable use of these technologies.

Chapter 10 shows different stages or versions of education. These are explained along the way which include Education 1.0, Education 2.0, Education 3.0, Education 4.0 and Education 5.0. It explains why innovation in education is imperatively required and how innovation is evolving day by day.

Chapter 11 intends to explain that despite the plethora of benefits of online classes, many loopholes and limitations cease to exist. The major issues can be caused by interruption and disturbance from unwanted students in the online classrooms. Their proposed work 'eScool' focuses on tackling these issues so that the online education experience is enhanced, and thus leads to improvement in students' productivity as well as efficiency.

Chapter 12 attempts to review the concept of self-directed learning by taking the basis of various theoretical frameworks that are quite significant in this context. Further, the study also provides a comprehensive review of the technology assisted online teaching learning, blended learning and lays down a conceptual model as an outcome of these reviews.

Chapter 13 takes Quality Assurance in Higher Education as an essential and integral part of various assessment and accreditation bodies and rankings across the world. It gives a complete picture of Education 5.0 which encompasses a paradigm shift from the conventional one-way monotonous teacher-centric learning to the two-way experiential student-centric learning. The chapter explores seamless integration of ten aspects of higher education in Education 5.0.

The work presented in the book will give some interesting insights to the readers.

Editor Biographies

Adarsh Garg: Professor at GL Bajaj Institute of Management and Research (GLBIMR), Gautam Buddh Nagar, Greater Noida and Visiting Professor at Delhi Technical University, Delhi. Prior to joining GLBIMR, she has worked with organizations like Galgotias University, WIPRO Tech, GE, IMT Ghaziabad, and Punjabi University, Patiala. She is currently supervising eight PhD students who have collectively published over 50 research papers in refereed international/national journals and conference proceedings. She is a member of various professional bodies like the Computer Society of India and ACM-Computer Science Teachers Association. She has 24 years of teaching, corporate and research experience with areas of interest in Business Analytics, Data Mining, Business Intelligence, Python, MIS, E-learning, and Project Management.

B V Babu: An acknowledged researcher and renowned academician, Dr B V Babu has over 36 years of teaching, research, consultancy, and administrative experience. He did his PhD at IIT Bombay. He is currently self employed as a Consultant on Quality Assurance in Higher Education and in the World Bank Sponsored Project HEQEP (Higher Education Quality Enhancement Program) since 2017. Previously, he was the Vice Chancellor of Graphic Era University-Dehradun (2016–2017), Vice Chancellor of Galgotias University-Greater Noida (2014–2016), Pro Vice Chancellor of DIT University-Dehradun (2013–2014) and founding Director of Institute of Engineering and Technology (IET) at JK Lakshmipat University-Jaipur (2011–2013). Prior to that, Dr. Babu served at BITS Pilani for 15 years (1996–2011), and held various administrative positions. Professor Babu is a distinguished academician and an acknowledged researcher with an H-index of 43 & i10-index of 118 and over 17,329 citations (as of April 2021). He is well known, internationally, for his algorithm, MODE (Multi Objective Differential Evolution) and its improved variants. Besides several highly-cited publications in international journals, he also has three well-accepted textbooks (published by Springer, Germany; Apple Academic Press, USA; and Oxford University Press, India) to his credit. He has supervised ten PhD candidates and is currently guiding two more PhD candidates. He is a member of various national and international academic and administrative committees.

Valentina E Balas is currently Full Professor in the Department of Automatics and Applied Software at the Faculty of Engineering, "Aurel Vlaicu" University of Arad, Romania. She holds a PhD in Applied Electronics and Telecommunications from the Polytechnic University of Timisoara. Dr Balas is author of more than 270 research papers in refereed journals and international conferences. Her research interests are in Intelligent Systems, Fuzzy Control, Soft Computing, Smart Sensors, Information Fusion, Modeling and Simulation. She is the Editor-in-Chief of *International Journal of Advanced Intelligence Paradigms* (IJAIP) and *International Journal of Computational Systems Engineering* (IJCSysE), a member of the editorial board of several national and international journals, and she is an expert evaluator for national, international projects and PhD theses. Dr Balas is the director of Intelligent Systems Research Centre in Aurel Vlaicu University of Arad and Director of the Department of International Relations, Programs and Projects at the same university. She served as General Chair of the International Workshop Soft Computing and Applications (SOFA) in eight editions (2005–2018) held in Romania and Hungary. Dr Balas has participated in many international conferences as Organizer, Honorary Chair, Session Chair and member of Steering, Advisory or International Program Committees. She is a member of EUSFLAT, SIAM and a Senior Member in the IEEE, member in TC – Fuzzy Systems (IEEE CIS), member in TC – Emergent Technologies (IEEE CIS), and member in TC – Soft Computing (IEEE SMCS). Dr Balas was past Vice-president (Awards) of IFSA International Fuzzy Systems Association Council (2013–2015) and is a Joint Secretary of the Governing Council of the Forum for Interdisciplinary Mathematics (FIM), a multidisciplinary academic body in India.

Contributors

Sapan Adhikari
Managing Director
SST.Pvt.Ltd
Attariya, Kailali, Nepal

Ishika Agarwal
Department of Computer Science
University of Engineering and Management
Kolkata, India

Naazneen Ahmed
Department of Computer Science
University of Engineering and Management
Kolkata, India

Neerja Aswale
Vishwakarma University
Pune, Maharashtra, India

B V Babu
Consultant on Quality Assurance in Higher Education
India

Valentina E Balas
Professor
Department of Automatics and Applied Software
Faculty of Engineering
University of Arad
Arad, Romania

Sunil Kumar Bharti
Department of Information Technology
Galgotias College of Engineering & Technology
Greater Noida, India

Ritika Chamaria
Department of Computer Science
University of Engineering and Management
Kolkata, India

Shrishti Choudhary
Department of Computer Science and Engineering
Sharda University
Greater Noida, India

Abhik Kumar De
Department of Computer Science and Engineering
Sharda University
Greater Noida, India

Abhishek Deupa
Department of Computer Science and Engineering
School of Engineering and Technology
Sharda University
Greater Noida, India

Jagjit Singh Dhatterwal
Department of Artificial Intelligence
 & Data Science
Koneru Lakshmaiah Education
 Foundation
Vaddeswaram, Andhra Pradesh,
 India

Adarsh Garg
GL Bajaj Institute of Management
 and Research
Greater Noida, India

Debanjana Ghosh
University of Engineering &
 Management
Kolkata, India

Subhajit Ghosh
Department of CSE
IMS Engineering College
Ghaziabad, India

Satyendra Gupta
Professor and Dean
School of Education
Galgotias University
G.B. Nagar, U.P. India

Svitlana Illinich
Associate Professor
Department of Social Technologies
Vinnytsia Institute and College of
 Open International University of
 Human Development
Vinnytsia Oblast, Ukraine

Kuldeep Singh Kaswan
School of Computing Science and
 Engineering
Galgotias University
Greater Noida, India

Ruqaiya Khanam
Department of Electronics and
 Communication Engineering
Center for Artificial Intelligence in
 Medicine
Imaging & Forensic
Sharda University
Greater Noida, India

Niranjan Kulkarni
Vishwakarma University
Pune, Maharashtra, India

P Pradeep Kumar
Galgotias University
Greater Noida, India

Srilekha Mukherjee
Department of Computer Science
University of Engineering and
 Management
Kolkata, India

Rudra Pratap Ojha
Department of Computer Science &
 Engineering
GL Bajaj Institute of Technology
 and Management
Greater Noida, India

Shraiyash Pandey
Department of Computer Science
 and Engineering
Sharda University
Greater Noida, India

Diptarka Paul
Department of Computer Science
University of Engineering and
 Management
Kolkata, India

Ravinder Rena
Faculty of Management Sciences
Durban University of
 Technology (DUT)
Durban, South Africa

Subham Sarangi
Department of Computer Science
University of Engineering and
 Management
Kolkata, India

Sarika Sharma
Symbiosis Institute of Computer
 Studies and Research
Symbiosis International (Deemed
 University)
Pune, India

Yamini Sharma
Department of Computer Science
University of Engineering and
 Management
Kolkata, India

Archana Singh
Vishwakarma University
Pune, Maharashtra, India

Rashmil Singh
Consultant, Higher Education
Georgia, USA

Ravi Kant Verma
Research Scholar
School of Education
Galgotias University
G.B. Nagar, U.P. India

Shreya Virani
Symbiosis Centre for Management
 Studies
Symbiosis International
 (Deemed University)
Pune, India

Acknowledgements

It is our pleasure to express our deep sense of gratitude to CRC Press, Taylor & Francis Group for providing us the opportunity to work on the project of editing this book, *Advances in Technological Innovations in Higher Education: Theory and Practices*. We would like to express our sense of gratification and contentment at the completion of this project. We express our gratitude from the bottom of our heart to all those who facilitated us in both direct and unintended ways to accomplish the task. First of all, we would like to thank the authors who have contributed to this book. We acknowledge, with sincere appreciation, the compassion of various authors at their respective institutions to carry out this work. We take this exclusive opportunity to express our sincere appreciation to Ms. Gabriella Williams, Editor, CRC Press, Taylor & Francis Group, for her sincere suggestions and kind patience during this project. We would like to thank our friends and faculty colleagues for the time they spared in helping us through the project. Special mention should be made of the timely help given by various reviewers during this project, though their names cannot be revealed here. The valuable suggestions they provided to the authors cannot be left un-noticed. We are enormously thankful to the reviewers for their backing during the process of evaluation. While writing, contributors referenced several books and journals; we take this opportunity to thank all those authors and publishers. We thank the production team of CRC Press, for encouraging and extending their full cooperation to complete this book. Last but not least we are thankful to the Almighty for guiding our direction.

Abbreviations

AR	Augmented Reality
VR	Virtual Reality
COVID-19	Coronavirus Disease 2019
NPTEL	National Programme on Technology Enhanced Learning
AI	Artificial Intelligence
ERP	Enterprise Resource Planning
NEP	National Education Policy
EMF	Electromotive Force
TV	Television
STEM	Science, Technology, Engineering, and Mathematics
BC	Before Christ
AD	Anno Domini
ICT	Information and Communication Technology
PCK	Pedagogical Content Knowledge
MOOC	Massive Open Online Course
LMS	Learning Management System
HBX	Harvard Business School Online
cMOOC	Connectivist MOOC
xMOOC	eXtended MOOC
MIT	Massachusetts Institute of Technology
SWAYAM	Study Webs of Active-Learning for Young Aspiring Minds
PET	Positron Emission Tomography
MRI	Magnetic Resonance Imaging
CT	Computed Tomography
ML	Machine Learning
CNN	Convolutional Neural Networks
SVM	Support Vector Machine
2D	Two-dimensional
3D	Three-dimensional
FC	Fully Connected Layer
GANs	Generalized Adversarial Networks
MAQ	Mobile Anonymous Question-Raising System

HMM	Hidden Markov Models
LFPC	Log Frequency Power Coefficients
ARHMM	Autoregressive Hidden Markov Model
Mel-LPC	Mel-frequency Scale
MFCC	Mel Frequency Cepstral Coefficient
LSTM	Long Short-term Memory Networks
MLP	Multi Perceptron Model
LRS	Learning Record Store
PLE	Personal Learning Environment
ADDIE	Analyze, Design, Develop, Implement and Evaluate
SPSS	Statistical Package for the Social Sciences
AIEd	AI in Education
UI	User Interface
DNN	Deep Neural Networks
HCML	Human-centered Machine learning
SNS	Somatic Nervous System/Simple Notification Service
IoT	Internet of Things
XR	Extended Reality
SDL	Self-directed Learning
IT	Information Technology
BL	Blended Learning
SMAC	Social Networks, Mobile Network, Analytics, and Cloud
OBE	Outcome-based education
GA	Graduate Attributes
PEO	Program Educational Objectives
PO	Program Outcomes
PSO	Program Specific Outcomes
CO	Course Outcomes
UNESCO	The United Nations Educational, Scientific and Cultural Organization

Chapter 1

Rehumanizing education in the age of technology

Subhajit Ghosh
Department of CSE, IMS Engineering College, Ghaziabad, India

1.1 INTRODUCTION

The Program Outcome in Outcome-Based Education states that the Program (B.Tech, MBBS, etc.) intends to build professionals (graduating students) capable of independent thinking. However, most faculties are often approached by students asking for notes from lectures, so that they can study and obtain good scores in their exams. At times, such notes contain only the essentials for exams, which students memorize and reproduce to secure good grades. Faculties too are complimented for good performance by their students. This is counterproductive to the end goal of outcome-based education. How can a system that fosters Learning by Rote, sorry Notes, achieve the goal of equipping students with the ability to think on their own? There is a need to reorient our education system to the demands of modern times.

There exists huge potential in technology-enabled learning. With the aid of technology these days we have been able to bring about many benefits that include better reachability, efficiency, cost-effectiveness, flexibility, scalability, and paperless education. Technology is a game-changer for education that can be used both in 'beneficial' and 'harmful' ways [1].

In the last few decades, education has undergone a sea change. Expectations from Higher Education sectors now include not limiting themselves only to teaching, doing research, or serving the community, but also innovating and bringing out transformation in the industrial processes. Universities must now gear themselves toward outcome-focused Nation-building activities that would solve such problems that would result in value creation.

While all of this is beneficial, the dominance of technology is accelerating a 'dehumanized' version of education. Reducing interactions between humans and shifting decision-making power from humans to technology can bring about undesirable consequences. Instead of a blind adoption of technology, we need to have a futuristic vision of education that puts humans at the forefront of the educational process [2,3].

DOI: 10.1201/9781003376699-1

The advent of artificial intelligence, automation, AR and VR, and other technological developments has been transforming modern education even pre-COVID-19. The pandemic compelled academic institutions to embrace digitalization. This produced varying levels of success, from the standpoint of quality and learners' performance, the motivation level of the students, and aspects of physical and emotional well-being. The lasting effect spans across age groups and is yet to be measured. Many experts opine that this is not the correct approach to strive for an education with a vision. Digital measures that were adopted by educational institutions exposed inadequacies and showed the value of human teaching in a physical capacity and in learning in the initial phases of education [3]. Table 1.1 depicts the evolution of education.

Table 1.1 Evolution of education

SNo	Category	Main features	Remarks
1	Education 1.0	In Education 1.0, students go to school to learn where teachers generally give them information and tell them about relevant books and videos and where they may be available	Still in vogue in many academic institutions
2	Education 2.0	Education 2.0 is a collaborative effort between educationists, policymakers, and researchers to look for innovative solutions to the most challenging problems. This had an influence on the process of teaching and on learning	The onset of innovative practices to address challenging problems and link education to benefit society
3	Education 3.0	Education 3.0 is based on the premise that content should be easily accessible. It is driven by self-motivation and lays emphasis on problem-solving and creativity	The Internet has facilitated easy access to content. Creative teaching practices can rev up the motivation level of the learner
4	Education 4.0	Education 4.0 is aligned with future trends, self-paced learning, and preserving human values	Aligned with Industry 4.0, self-paced learning by doing NPTEL or Coursera courses online
5	Education 5.0	Education 5.0 is about transforming the present education system to an outcome-based system and bringing out a humane and compassionate element in the educational process	Work-in-progress; desirable

For educational practices to fructify, a systematic approach coupled with a vision is necessary, along with clarity of what it is exactly that we want to achieve with it. Industry 4.0 builds cyber-physical systems driven by automation, which requires skilling to deal with machines. The Fifth Industrial Revolution i.e., Industry 5.0, requires 'rehumanizing' manufacturing and service processes. This implies humans play an interactive role and there exists effective and efficient cooperation between humans and machines. If education is to have a lasting impact, there is a need to align it with the Fifth Industrial Revolution that is being talked about these days. We need Education 5.0.

1.2 DIGITALIZATION OF EDUCATION AND SALIENT FEATURES OF EDUCATION 5.0

Education 5.0 aims to change the present education system to action/outcome-based systems. Outcome-based education has been the focus for some time now. Though there have been considerable attempts to provide need-based training and facilitate industry-related projects in the curricula, the quality of graduates produced every year continues to deteriorate. The actual needs of businesses simply can't be fulfilled by the quality of the graduates coming out of most educational institutions.

Education 5.0 strives to put education on a realistic plane, equipped with a clinical understanding of the environment and theoretical conceptual knowledge. Students ought to be provided an experience of experiential learning for better outcomes.

Education 5.0 incorporates the following:

- Human related – This would emphasize identifying skills that only humans possess. The ability to create and think critically, skills in designing and analytical skills, compassion, innovativeness are unique in humans.
- Comprehensive education – Education should take into account developments in the world, and labor market and try to impart necessary skills using the best approach towards this end.
- Curriculum – All stakeholders should be engaged in curriculum development and implementation.
- Right balance – Education must impart the importance of maintaining good physical and mental health. Excessive technology can be harmful and the consequences of overuse should be emphasized.
- Lifelong pursuit of knowledge – Education should develop students to inculcate a spirit of continuous learning in their career.
- Personalized learning – Self-paced learning to be encouraged commensurate with the level of the student.

Education 5.0 lays more emphasis on humans, instead of technology. It focuses on the outcomes that need to be achieved by humans from a particular learning experience. This does not lay emphasis on improving infrastructure and connectivity or developing digital tools and platforms. It is about preparing intellectually and emotionally strong individuals, backed by appropriate strategic, methodological, and pedagogical approaches. It strives to bring motivation, creativity, and joy of learning back to learners. Digital equipment, infrastructure, and platforms are enablers towards the realization of the goal [3].

Technology is a tool for education that can be used rightly or wrongly. Digitization of outdated content and ineffective approaches is not a solution. Furthermore, some of the non-digital approaches that currently prove to be effective may lose their effectiveness in a digitized form. If applied wrongly, this tool may thus do more harm than good, so we had better do it right [1].

1.3 ARTIFICIAL INTELLIGENCE FOR PERSONALIZED LEARNING

We are witnessing increased enthusiasm regarding the use of AI in the classroom. China has already stolen a march, and many developed nations too are pursuing the same. Webinars/conferences are upbeat on the potential benefits that the use of AI can usher in for our students. Here one is not dwelling on the advantages of AI use for imparting education, which is unquestionable but trying to make sense of what an AI classroom would look like and the challenges of implementing it in the country. **An AI-based education (smart class) would equip a teacher/computer to provide personalized care for a group of students (assume 60). The Teacher/Robot/Intelligent Instructional System in the class would be able to gauge the level of the student, and his/her competency and then accordingly tailor the teaching methodology and deliver the content for the most optimal outcome. This ought to epitomize perfection in education.**

No two students are exactly the same. Some are visual learners, while some are hands-on. Some ask questions and some are a bit shy. For this reason, individualized learning is critical to each student's own educational experience.

Artificial intelligence is making personalized learning possible. At Carnegie Learning, for example, sophisticated AI technology adapts to each student's level of learning in real time. The platform provides personalized feedback, assessments, and guidance.

For parents, AI uses predictive analytics to forecast where a child is headed based on their individual performance in the program. This gives parents and educators a chance to intervene if a student is headed off track.

Now let us try to delve into the requisites to bring about such a smart class. Obviously, every student must have a computer. A few years back efforts to produce a 'Simputer' in India (a simple computer costing roughly 7,000 rupees) were heard of but no news thereafter accompanied the initial hype leading us to conclude that the project didn't prove to be much of a success. Besides these, we need good educational content (to be prepared by teachers), software developers (AI specialists and programmers), database administration (advanced servers, software, and database administrator), and good Internet connectivity. Creating this backbone would be a costly affair. In a country where the outlay for education is minimal, the dream of a truly smart class system seems more like a pipe dream. It may even prove to be a messy affair if implemented on a nationwide scale. Ever since software has made inroads into educational systems (ERP/Learning Systems, etc.) one has witnessed that faculties and teachers are engaged more in data entry and analysis rather than preparing lessons for taking their classes effectively. One hopes that a detailed study about the pros and cons of going smart in the classroom is made and a decision is taken about the level at which AI needs to be introduced in the curricula.

1.4 EMPHASIS ON EDUCATION 5.0

Education 5.0 requires a new outlook to address essential issues. The elements that need to be addressed are:

- Strategy: Outline the purpose and objectives as per Education 5.0
- Learning environment: Here we need to create a learning environment that addresses strategic objective issues related to design thinking and collaborative problem-solving, generate team spirit and risk-taking behavior as well as exhibit a multidisciplinary-oriented, experimental approach
- Delivery mechanisms: This would require the identification of tools that can meet the objectives of the Strategy element. In this case, an emphasis on technology as a delivery mechanism can be explored
- Collaboration: This would entail advocating practices that is beyond the normal collaboration of the institute so as to involve communities, and innovate to create an ecosystem of learning that would involve the main stakeholders
- Content: Identify, develop, and use content that is in line with the elements of strategy, i.e., a good mix of technical as well as aspects of non-technicality that includes questions about diversity, social inclusion, ethics, sustainability, etc.);
- Assessment and recognition: This would require developing established and off-the-cuff ways of evaluation and methods of recognition for Education 5.0; and
- Quality Assurance: This would require the development of appropriate criteria for ensuring quality in Education 5.0. This implies that

there is a need for monitoring quality on a continual basis. This has to look into aspects of the learner and the society and not merely the perspectives of the market and the employability requirements of the company.

1.5 INNOVATION IN TEACHING PRACTICES FOR EDUCATION 5.0

Institutions should develop multiple modes of engagement that promote greater active learning and enhanced access and provide flexibility to all students [4]. Innovations in teaching **encompass teachers reinventing their delivery methodologies by using several tools to make learning adaptive, inclusive, and dynamic.**

Many of the ideas regarding these challenges have been addressed in NEP 2020 such as Outcome-Based Education, Slow and Fast Learners, Project-Based Learning, Academic Bank of Credit, Blended mode of learning, and others.

The newest and the best technologies in the class shouldn't be synonymous with innovative teaching strategies. Only those recent teaching methodologies that improves the academic outcomes and look into aspects of real problems that promote non-discriminatory learning should be described as innovative.

In essence making use of novel teaching in the classroom is an acceptance that teaching practices can be improved upon further. It seconds growth and development, the expectations we have for our students. Needless to say, innovative teaching begins with a mindset to grow. Identification of areas for improving is made. We have to give sufficient time to research and evolve better methods to impart lessons to the students. Creation of novel or adaption of methods in practice is done. Risks are taken. Failure would necessitate we try once more. We keep on repeating and improving upon our mistakes doggedly thereby facilitating the establishment of innovation and help the class to think out-of-the-box thereby building competency in students greatly.

There is an imminent need to introduce innovation in teaching that would improve student outcomes. Such techniques attempt to engage the students actively towards learning that would improve the outcomes of the learning. It has been observed that students who are participative in their classes have higher attendance and usually are the better performers.

Overall, such efforts ought to be student-centric. Which approach is more beneficial for students to acquire knowledge during a one-hour class: passive listening while seated or active engagement through questioning, collaboration, and problem-solving?

Here are some alternative ways teachers can enhance student engagement and academic outcomes by incorporating innovative teaching strategies in their classrooms.

1.5.1 Flip the classroom

In a flipped classroom, the traditional roles of students and teachers are reversed. Instead of receiving lectures in class, students are assigned to review the lecture material at home. Classroom time is then dedicated to collaborative activities such as projects, problem-solving, and assignments. The coursework typically assigned as homework is now completed in class. This approach fosters peer-to-peer collaboration, allowing students to work together on group projects, engage in debates, and practice their skills. In the flipped classroom model, teachers take on a more flexible role. Rather than being the central focus, they provide personalized assistance and guidance to individual students and student groups as they work on their assignments.

1.5.2 Personalized learning

Personalized learning focuses on tailoring the education process to the specific needs, preferences, and abilities of each student. Instead of using a uniform approach for the entire class, teachers take into account the strengths and characteristics of individual students to enhance their chances of success. This personalized approach is akin to the customization we experience with various online tools, where algorithms adapt the online content to align with our interests. When users visit a website, they are presented with content that is most relevant to them based on their browsing history and searches, which may differ from what other users see. Personalized learning strives to create a customized learning experience by employing methods and techniques that are best suited for each student. While the specific individualized experiences may vary, the ultimate objective remains the same, which is to achieve mastery of the subject or meet the appropriate standards for that particular level. These diverse approaches can be likened to different paths that lead to the same destination.

1.5.3 Project-based learning

Project-based learning (PBL) is an educational approach that focuses on students actively engaging in a project or an extended task that aims to address a real-world problem or challenge. It involves students taking responsibility for their own learning by investigating and exploring the topic, conducting research, and developing a solution or product. One of the key benefits of project-based learning is that it promotes the development of essential skills such as critical thinking, problem-solving, research, and collaboration. By working on a project, students are encouraged to think critically and creatively to come up with innovative solutions. They learn how to apply their knowledge and skills in a practical and meaningful way, moving beyond rote memorization and gaining a deeper understanding of the subject matter.

In project-based learning, the teacher's role shifts from being the sole source of knowledge to that of a facilitator or guide. The teacher provides guidance and support to students throughout the project, helping them with research, offering feedback, and facilitating collaboration among students. This allows students to become more independent and take ownership of their learning process.

Furthermore, project-based learning is an effective method for fostering student engagement and motivation. By working on a project that has real-world relevance, students see the purpose and value of their learning, which can increase their motivation to actively participate and produce high-quality work.

Overall, project-based learning provides a dynamic and interactive learning experience that encourages students to think critically, solve problems, work collaboratively, and apply their knowledge in meaningful ways. It promotes a deeper understanding of the subject matter and equips students with essential skills for their future endeavors.

1.5.4 Inquiry-based learning

Inquiry-based learning is an educational approach that focuses on fostering problem-solving skills and critical thinking abilities in students. Rather than relying on traditional lecture-style teaching, the teacher plays the role of a facilitator who poses questions, scenarios, and problems to the students.

The process typically begins with the teacher presenting a thought-provoking question or a real-world problem to the class. Students are encouraged to explore and investigate the topic independently or in groups. They gather information, conduct research, and analyze data to formulate their answers and develop a deeper understanding of the subject matter.

Once students have gathered their findings, they have the opportunity to present their conclusions, supported by evidence, to the rest of the class. This presentation allows students to articulate their thoughts, communicate their ideas effectively, and build their presentation skills.

Furthermore, during the presentations, other students have the chance to ask questions, provide feedback, and engage in discussions. This interactive aspect of inquiry-based learning encourages students to critically evaluate their peers' work, consider alternative perspectives, and refine their own understanding.

By engaging in this iterative process of researching, presenting, and discussing, students not only expand their knowledge but also enhance their problem-solving abilities, analytical thinking, and communication skills. They learn to evaluate and synthesize information, draw logical conclusions, and articulate their thoughts effectively.

Inquiry-based learning empowers students to take ownership of their learning, encourages curiosity, and fosters a deeper engagement with the subject matter. It promotes an active learning environment where students actively participate, collaborate, and construct knowledge, rather than passively receiving information.

Overall, inquiry-based learning provides a student-centered approach that nurtures critical thinking, problem-solving, and effective communication skills, which are essential for success in both academic and real-world settings.

1.5.5 Ask open-ended questions

Many students, particularly those who excel academically, often rely heavily on textbook solutions. Over time, they may develop the belief that there are only definitive right or wrong answers. However, in reality, most questions do not have absolute truths.

In today's polarized public sphere, it is crucial for students to cultivate conversational skills and empathy. They should learn how to effectively communicate and collaborate with others. By prioritizing open-ended questions, teachers can foster lively discussions in the classroom. This approach encourages students to integrate various types of information they have learned or experienced, allowing them to construct coherent arguments. As a result, students not only discover their own perspectives but also learn how to express themselves effectively.

1.5.6 Peer teaching

Encouraging students to explain or teach others is an effective way to enhance their competency. Within the subject being taught, students should be given the freedom to select an area of interest. The teacher can facilitate this process by allowing students to independently research their chosen topic and create a presentation about it. Allotting class time for students to present to their peers enables them to teach and share their knowledge with others. This peer teaching approach not only helps students develop independent study skills and improve their presentation abilities, but also fosters self-confidence.

1.5.7 Feedback

Constructive feedback plays a vital role in the learning process, and it is essential for students to develop the skills to give and receive feedback effectively. Teachers should establish a system that allows students to provide feedback. In virtual classrooms, feedback tools such as polling or emojis can be utilized to gather quick feedback from students. Moreover, encouraging students to elaborate on their feedback and initiating discussions among students with different perspectives can further enhance the learning experience.

1.5.8 Blended learning

Blended learning combines in-person and online instruction, empowering students with greater autonomy in choosing when, where, and how they learn. It offers a unique blend of traditional classroom experiences and digital resources, opening up diverse learning opportunities. Technology plays a vital role in this approach, mirroring the importance of technology in students' lives beyond the classroom. The flexible nature of blended learning allows students to customize their learning experience, whether it involves watching online lectures at home and participating in collaborative activities with peers or opting for virtual classes with lectures while completing independent homework assignments.

1.5.9 Active learning

Several of the learning strategies we talked about revolve around active learning. Active learning approaches involve stimulating students to engage in discussions, contribute ideas, participate actively, conduct investigations, and generate new content. By questioning students, encouraging problem-solving, and fostering critical thinking, active learning prompts them to be actively involved in the classroom. When students participate in their own learning process, they tend to have a higher likelihood of succeeding in the class.

1.5.10 Jigsaws

Jigsaw activities serve as an interactive learning approach, offering students the chance to teach their peers, aligning with Seneca's belief that 'while we teach, we learn'. Explaining concepts to others is often considered the most effective way to truly grasp them, known as the protege effect. By using jigsaw activities, students are divided into groups and given different pieces of information. They must then understand their assigned information well enough to explain it to others. The students then rotate between groups, sharing their knowledge until each group has acquired a comprehensive understanding of the entire topic, akin to completing a puzzle.

In recent years, the transition from physical classrooms to virtual learning has accelerated, with more students gaining proficiency in digital tools. Even before the pandemic, there was a significant increase in enrollment in virtual academies, catering to millions of students annually. While some institutions have reopened since then, it is unlikely that schools and colleges will completely abandon their digital experiences. Digital learning offers students increased flexibility, granting them greater access to teachers and a wider range of classes. It also empowers students to take more control over their own education. As Plato wrote, 'our need will be the real creator', reflecting the idea that necessity drives innovation. Innovative teaching strategies, once considered niche practices of a few adventurous educators, are now becoming more commonplace as institutions strive to address learning gaps and adapt to our new reality [5–7].

In the days ahead, we are likely to witness a spurt in blended learning, hybrid learning, and innovative initiatives to meet the challenges of modern students. This is not only relevant for students but also at the workplace which too faces these challenges and attempts to find its own hybrid learning experiences. These strategies are used to inspire creativity and success in the classroom. There is a need for change and through it, we are bound to succeed or become a failure. However, failing is ok. One of the most important lessons we teach our students is that they need to try and if they fail, then that's okay. Failing is okay so far, we take lessons from that and try again.

Though these strategies seem like we are taking a big leap into something new, we don't have to apply them to our entire teaching strategy. Think of how you can use one or the other for a specific lesson. Maybe some subjects lend themselves to a project-based learning exercise while others benefit from simply asking open-ended questions. As an illustration, the project-based learning approach can be used in certain portions of the subject 'Artificial Intelligence', whereas peer-based learning can be used for a subject like 'Rural Development'.

Many teachers are going through a similar experience while innovating classroom teaching to make it more effective. There are some fantastic examples online that one can use as source material for classroom experiments.

The teacher can try out different technologies like recording video lectures or using virtual classrooms when appropriate to venture into the digital experience. **Maybe even having students create their own videos to teach and inform other students – our students are already creating videos with their friends, so maybe one can leverage their excitement and put it to good academic use.**

As long as we're innovating, we are growing! Give it a go, it's always an exciting time to be in the classroom. It's especially fulfilling now that many are looking to introduce innovative teaching strategies as solutions to the challenges students face today.

1.6 ROLE OF GOVERNMENT IN EDUCATION 5.0

Governments play a crucial role in promoting the principles of Education 5.0. They can take several measures, including the following:

- Instead of prioritizing technology, governments should focus on achieving desired learning outcomes for individuals during the transformation process

- Governments should support initiatives that identify and promote successful practices for each element of Education 5.0.
- Governments should encourage the development of common guidelines for educational institutions and training providers to design and implement Education 5.0 effectively, and facilitate their widespread adoption.
- Governments should take steps to monitor, analyze, and prevent negative practices in digital education, such as using outdated course content, which can result in reduced motivation, performance, and health issues for learners.
- Governments should establish guidelines for educational institutions and learners concerning data protection, privacy, and other ethical considerations related to digital technologies.
- Governments should pay special attention to creating a healthy environment in schools due to the growing use of technology. They should promote initiatives to reduce electromagnetic field (EMF) radiation in schools and colleges by implementing eco-friendly wireless routers and ethernet connections. Additionally, they should encourage physical activity among learners and foster mindfulness about screen time.

By implementing these measures, governments can effectively promote Education 5.0 and ensure its positive impact on learners and education providers.

1.7 VALUE-BASED AND MENTAL DEVELOPMENT FOR EDUCATION 5.0

The issue of youth suicide has become a prominent concern in the Indian academic system, demanding urgent attention and effective preventive measures [8]. The increasing number of suicides among young people calls for a deeper understanding of the underlying factors contributing to this distressing trend. One significant factor is the intense competition and limited opportunities in the education sector. Admission into prestigious educational institutions, particularly in the National Capital of India, has become extremely challenging, with high cut-offs percentages ranging from 99 to 100%. This intense competition can lead to depression and despair among students who are unable to secure admission to their desired colleges despite having reasonably good grades. Moreover, pursuing a desired course of study has become financially burdensome, with many students burdened by educational loans. This financial strain further exacerbates the situation when students face additional challenges or difficulties within their chosen courses [9].

The rigid nature of our educational system also contributes to the problem. To address this, it is crucial to revamp the evaluation system and incorporate a 'fitness for the course index' (Score) that assesses a student's aptitude and innate skills for their chosen professional course within the first two semesters. This early assessment would help identify students who may be unsuited for the course and guide them towards alternative studies or activities, preventing them from enduring unnecessary struggles and potential mental health issues.

Overall, it is imperative to recognize the gravity of the situation and implement measures that prioritize the well-being and mental health of students.

There is also a need to emphasize on right values to be imparted to our students in their learning program. Society is riddled with crimes and undesirable acts being committed most often by the educated. The need for a value-based education has never been greater than now.

1.8 PSYCHOLOGICAL DAMAGE TO A LEARNER AND OTHER DEMERITS OF TECHNOLOGY

To illustrate the above, let me narrate an incident that transpired some years back. One day, my wife said that during the day she had received a WhatsApp message from the class teacher of our son. It showed him in a picture as one of the three students lined up for punishment. The picture was similar to the ones we see in movies or TV and in newspapers, where criminals are made to line up and photographs are taken of them where humiliation is writ large on their faces. The crime the three high school standard boys committed was not doing their homework, which emboldened the subject teacher to take a snap and forward it to the class teacher, who WhatsApp-ed that same picture to my wife. When my wife strongly objected to this way of punishing a child, reasoning that such issues should be resolved at school, the class teacher informed us that they keep sending such pictures of insincere students being punished similarly, and hardly any parents have criticized this method. One doesn't know how far this is true about 'accepting parents' but this practice is a blatant misuse of technology, where groups of classes are formed on WhatsApp, and activities and information and note sharing happen therein. Imagine if the teacher posts and the child see his picture flashed in his class group on WhatsApp!!! It may damage his morale beyond measure. Thankfully, this didn't happen in the case of our child.

The omnipresence of technology, including artificial intelligence, machine learning, robotics, and automation, has brought about several negative impacts on various aspects of our lives.

Ability to write – One such consequence is observed in our ability to write effectively. With the widespread use of abbreviations and shortcuts like 'u' instead of 'you' in SMS messages and WhatsApp conversations, this trend has now made its way into students' answer scripts. As a result, individuals who adopt these abbreviated forms fail to fully grasp the beauty of language, thereby impairing their writing skills and diminishing their appreciation for the intricacies of the language.

Imagination and Wonderment – Einstein once said: *"Imagination is more important than Knowledge."* Looking three to four decades back, there wasn't any TV around and our world was full of stories and imagination and fantasy ruled whenever we thought of them. We would visualize our heroes like 'The Three Musketeers' and 'Robin Hood' and their acts of valor through the power of our imagination. The sense of wonderment that filled the life of a child immortalized by Satyajit Ray in the classic train sighting sequence by Apu and Durga in the film Pather Panchali has disappeared from the lives of children these days.

1.9 ASSESSMENT OF THE TEACHING AND PEDAGOGY

This relates to the issue of student feedback. One has generally found that student feedback consists of numeric numbers indicative of a degree of acceptability of the faculty by the student. Except for a general opinion that may be derived, this feedback doesn't help in the qualitative improvement of teaching by the faculty. Faculty with poor feedback would only be cautious and try to improve their acceptability amongst the students by working extra hard. The ones with good feedback become generally complacent without feedback on some of the points where they may improve further.

Rather than numeric feedback, descriptive feedback on the faculty by the students would better serve our objective of improving faculty teaching. As an example, if a student is asked to write (in words) **three strong points and three weaknesses of teaching by the faculty**, we would have a large collection of about 120 strong points and 120 weaknesses of a faculty (assuming a class strength of 40), and even if we can summarize ten important points about strength and weakness, we will have arrived at a better estimate of the quality of teaching. Moreover, if a student ponders about 5–10 minutes on the teaching of a particular faculty to write out these six sentences rather than assigning values quickly to pre-set parameters, it would probably result in identifying the strengths and weaknesses of the teaching and the pedagogical process in a clearer way.

1.10 CONCLUSION

Education 5.0 prioritizes the human element over technology and utilizes technology as a tool to enhance value and effectiveness. Rather than focusing on the quantity of technology used, Education 5.0 emphasizes making thoughtful and responsible choices while considering the broader context. Privacy, ethics, safety, and technological mindfulness are given special attention in Education 5.0. This approach to education transformation takes a comprehensive view, encompassing all vital aspects. Additionally, Education 5.0 encourages collaboration among various key stakeholders, including governments, educational institutions, industry, support structures, the wider community, and, most significantly, learners themselves. By uniting in this manner, we can ensure that we proceed in the right direction, paving the way for a promising and sustainable future for both current and future generations.

REFERENCES

[1] European Commission, Executive Agency for Small and Medium-sized Enterprises. (2020). Skills for industry curriculum guidelines 4.0 – Future-proof education and training for manufacturing in Europe: final report, Publications Office. https://data.europa.eu/doi/10.2826/097323.

[2] Jonathan, E. (2019). Education 5.0 – towards problem-solving and value creation – Ministry of Higher and Tertiary Education, Science and Technology Development (mhtestd.gov.zw).

[3] Dervojeda, K. (2021). Education 5.0: Rehumanising Education in the Age of Machines (linkedin.com).

[4] Karbhari, V. M. (2022). Time to re-envision higher education, The Times of India Higher Education, March 2022, Page 4–5.

[5] Thompson, S. (2021). Innovative Teaching Strategies, Innovative Teaching Strategies | Kaltura.

[6] Burkhalter, M. (2021). How AI in the classroom is supporting education (perle.com).

[7] Braun, B. O. (2013). Innovative methods in Engineering Education, Innovative Methods in Engineering Education – ppt download (slideplayer.com).

[8] Shetty, H. (2022). Opinion | First Sushant Singh Rajput and now Tunisha Sharma – How Irresponsible has Indian Media Become? (news18.com).

[9] Ghosh, S. (2012). Suicide Among Indian Students? Why?? by Subhajit Ghosh (boloji.com).

Chapter 2

Transitioning pedagogies in evolving India

Critical analysis of skills, knowledge, and wisdom with respect to implementation of NEP 2023

Neerja Aswale[1], Niranjan Kulkarni[1], Rashmil Singh[2], and Archana Singh[1]

[1]Vishwakarma University, Pune, Maharashtra, India
[2]Consultant, Higher Education, Georgia, USA

2.1 INTRODUCTION

The work in [1] has provided an insight towards the NEP 2023 which introduces the modern Indian education system to be inclusive, equitable, and holistic. The focus of NEP 2023 is to evade rote learning and memorizing techniques and welcoming creativity, critical thinking and problem solving skill enhancement through varied teaching pedagogies. India, a country noted for its diverse population and rich cultural past, is witnessing a dynamic shift in its education system. As the country recognizes the need to prepare its students for the challenges and opportunities of the 21st century, the expanding landscape of Indian education is witnessing a shift in pedagogical approaches. This shift in pedagogies is being driven by a rising understanding that traditional methods of teaching and learning may not be adequate to provide students with the skills and competences needed in a fast-changing environment. Historically, the Indian education system was characterised by a teacher-centred approach that emphasised rote memorization and passive learning. However, as globalisation, technological improvements, and the needs of a knowledge-based economy become more prevalent, there is a growing recognition that students must develop critical thinking, problem-solving, creativity, cooperation, and communication skills. As a result, there is a movement towards learner-centred pedagogies that emphasise active engagement, inquiry-based learning, and knowledge application in real-world contexts. The realization of the need to stimulate student creativity and entrepreneurship is one of the primary drivers of this change. India's ambitions to become a global powerhouse for innovation and technology necessitate an education system that fosters creativity, risk-taking, and an entrepreneurial spirit. To this end, pedagogical approaches are being adapted to emphasise learning experientially, project-based learning and cross-disciplinary education in the areas of STEM science, technology, engineering and mathematics.

Additionally, the quick development of digital technologies has created new opportunities for Indian education. It is now simpler to reach students in rural locations and give them a decent education because of the proliferation of inexpensive cell phones, greater internet connectivity, and the availability of digital content. The popularity of blended learning methods, which supplement conventional classroom education with online tools and platforms, is being driven by the digital revolution. Additionally, it enables students to have personalized educational experiences based on their unique requirements and interests. Additionally, inclusivity and equity are key factors in the emerging pedagogies in India. An effort is being made to make sure that education is available to everyone while also taking into account the diversity of learners in terms of their talents, backgrounds, and learning styles. To meet the unique needs of kids, this involves fostering inclusive classrooms, adaptive learning tools, and customised education. With a move towards learner-centred pedagogies, the incorporation of technology, and an emphasis on encouraging innovation and entrepreneurship, India's educational system is going through a transformational period. This change acknowledges the need to provide students with the competencies, knowledge, and skills required for success in a world that is rapidly changing. India is establishing the groundwork for a future-ready educational system that equips its youth to succeed in the 21st century by adopting these changing pedagogical approaches.

2.2 THE AIM OF THE CHAPTER

1. To analyse the timeline of the Indian Education System from ancient to modern times.
2. To identify the relevance of skills, knowledge and wisdom-based education in NEP 2020 implementation
3. To identify the teaching pedagogies to be utilized in NEP 2023 implementation in terms of skills, knowledge and wisdom.

2.3 LITERATURE REVIEW

2.3.1 The evolution of teaching pedagogy in pre-British, medieval, and modern India

Education has always been important in moulding civilizations, and teaching style has developed over time to accommodate the changing requirements of students. The evolution of teaching pedagogy in India may be traced back to the pre-British, medieval, and modern eras, each of which was impacted by diverse circumstances and cultural settings. This essay explores the evolution of teaching pedagogy in India across these three periods and highlights the key characteristics and transformations that occurred.

2.3.2 Pre-British India

In pre-British India, education was deeply rooted in the guru-shishya (teacher-student) tradition, which emphasised a personalized and holistic approach to learning. The guru, or teacher, held a revered position and imparted knowledge to their disciples through oral instruction and experiential learning. This pedagogy focused on values, character development, and practical skills relevant to everyday life.

The gurukul system was prevalent during this period, where students resided with their gurus and imbibed knowledge not only through formal instruction but also by observing and emulating their teachers' behaviour. Education encompassed a wide range of subjects, including philosophy, mathematics, astronomy, medicine, music, and more. Learning was primarily conducted in small groups, fostering individual attention and mentorship.

2.3.3 Medieval India

The medieval period in India witnessed significant changes in teaching pedagogy due to the influence of Islamic scholars and rulers. Madrasas emerged as centres of learning, primarily focused on religious education and the study of Islamic texts. The curriculum revolved around theology, law, and Arabic language studies. The teaching pedagogy in medieval India emphasised rote memorization and recitation of texts, with teachers playing a central role in transmitting knowledge to students. Instruction was often delivered in a lecture format, limiting student engagement and critical thinking. However, some educational institutions, such as Nalanda and Takshashila, continued to follow a more holistic approach, offering diverse subjects and encouraging scholarly discussions and debates.

2.3.4 Modern India

The British invasion of India resulted in a significant shift to pedagogical ideology. They introduced a formalized education system, modelled on Western ideals, with an emphasis on standardized curricula, examinations, and institutionalization. English became the medium of instruction, leading to a gradual decline in the usage of indigenous languages. The British education system focused on subjects such as science, mathematics, and literature while neglecting traditional knowledge systems. Classrooms became more structured, with teachers adopting a didactic approach, transmitting knowledge through lectures and textbooks. This method of teaching emphasised rote learning and examination-based assessment, limiting critical thinking and creative expression. The limits of the British educational system in India have recently come to light more and more. A more holistic and student-centred approach to teaching and learning is being promoted. Progressive pedagogical methods, such as experiential learning, collaborative activities, and project-based learning, are gaining popularity. Technology has also played a significant role in transforming teaching pedagogy, enabling access to a vast range of educational resources and facilitating interactive and personalized learning experiences.

The first Objective of the paper is answered through the **Timeline of Ancient, Medieval and Modern Education:** [2] Authors have depicted the time line for the education system from Vedic to Modern Education. From around the 5th to 12th centuries BC the Ancient Education evolved which comprised Vedic and Buddhist Studies which taught Sanskrit and Pali simultaneously. There were Universities to teach Vedic and Buddhist Education which comprised Takshashila University, Nalanda University, Vallabhi University, Sharda Peeth, Mithila University, Pushpagiri Vihara, Vikramshila, etc. The renowned teachers from the ancient period were Panini, Chanakya, Kumardata, Vasubandhu who taught subjects ranging from Indian Governance to Tibetan Buddhist Studies. The education in this era included Vedas, Brahmanas, Upanishads, and Dharmasutras. The teaching was in Gurukul away from parents being with nature so as to connect with oneself. The ancient period education included studies towards development of students not only in terms of external body but also internal and mental health. The work in [3] depicted in the article of LinkedIn that ancient times learning was holistic aiming towards trust ethics, humility, discipline and self-reliance wherein the emphasis was on fostering the mental, physical and intellectual growth. The work in [4] highlighted in his essay on medieval education, which lasted from the 10th century A.D. until the middle of the 18th century, i.e., before British rule. The structure of education in medieval and early modern was influenced by Muslim Educational Models which had Maktabs and Madarsas to teach primary education like basic reading writing and Islamic prayers. This also included advance language skills. After arriving in India, the Mughal emperors established their power. The great interplay between Indian and Islamic traditions in all areas of knowledge, including theology, religion, philosophy, fine arts, painting, architecture, mathematics, medicine, and astronomy, during that time gave education a new depth [5]. With the advent of British control in the 17th century and the erroneous policies of the then rulers, India lost practically all of her intellectual wealth. In the year 1830, Thomas Macaulay developed the present system, which included English as the principal language and also included Mathematics and Science, whereas disciplines such as philosophy and metaphysics were viewed as superfluous. The link to nature and the close relationship between the teacher and student were broken, and the training was only provided in classrooms. This gave rise to rote learning and memorizing techniques and which avoided the problem solving and critical thinking skills which had been a matter of concern for the Indian education system [6]. Lately, the Indian Government has enacted intermittently varied educational policies starting from Sarva Siksha Abhyan in 2001, Rashtriya Madhyamik Siksha Abhiyan and Right to Education in 2009. Finally, the New Education Policy which had been drafted in various years i.e., 1986, 1992 and finally was implemented in July 2020 and implementation in 2023. The NEP 2020 has the goal of providing excellent education to all students. It also emphasises the necessity of teaching a broader range of courses, such as the arts, athletics, and vocational skills, in addition to conventional academic subjects.

The Second Objective of the chapter emphasises skill-, knowledge- and wisdom-based teaching pedagogies depicted below:

A. Relevance of skills-based education: Need for implementing NEP 2020

Since graduates join the workforce after at least 12 years, if not more, of formal education intended to prepare them to manage their financial and social lives as adults, educators are baffled by the absence of "life skills" in the classroom [7]. Companies (and now the government) are undertaking a massive job that should need more than 12 years of formal education. A seemingly limitless and frequently overlapping list of life skills is offered, with the study's recommendations summarised in four points: (a) learning to know; (b) learning to do (critical thinking, problem solving, communication and collaboration, creativity and innovation, information, media, and technology literacy, and information and communication technology [ICT] literacy); and (c) being a learner (social and intercultural skills, personal responsibility, self-regulation and initiative, perception skills, metacognitive skills, entrepreneurial thinking skills, learning to learn, and lifelong learning pursuits) [8]. In response to recent concerns from Indian business on the "employability of school and university graduates," this study investigates the role of pedagogy in building life skills and how they might be incorporated into school/university curriculum. Life skills have been included in the school curriculum recently, with a focus on the value of inquiry and teamwork across all topics taught in the classroom [9]. The importance of sustainability education at universities is generally understood, and it is being introduced increasingly often. However, education for sustainable development and education for sustainability reflect differing degrees of curricular change, but attaining sustainable education will need much more change. As more people argue for a variety of analytical and context-related abilities to be cultivated in students, a transformational pedagogy supports and contributes to the amount of the shift. Teaching strategies must put more emphasis on learning processes than knowledge accumulation in order to operationalize sustainability-related education and produce graduates who can improvise, adapt, innovate, and be creative. These skills are covered by the problem-based learning (PBL) pedagogy, which offers students the chance to learn how to think, especially "how to think" as opposed to "what to think," and maybe within the context of sustainability. As a result, it is critical to pinpoint the similarities between PBL, sustainable education, and transformative learning. The task is to change our pedagogy across all disciplines so that academics and students are encouraged to think critically, which is a fundamental link in this discussion. This article goes into greater detail on these ideas and makes the case that the important component of sustainability education is the growth of thinking [10]. NEP 2020 aims to provide more flexible and diverse educational routes, allowing students to pursue courses of interest that best match their strengths and ambitions. Infusing skill development into the school system is the solution to increasing the effectiveness of the education system.

B. Relevance of wisdom-based education: Need for implementing NEP 2020

The work emphasises that the theory of the balance of wisdom was described and applied in the environment of education. The essay begins by discussing why intelligence-related abilities are a necessary but insufficient foundation for teaching. Second, the paper briefly examines several wisdom theories. Third, the study introduces the balance theory of wisdom, which defines wisdom as the use of value-mediated implicit and explicit knowledge to promote the common good through a balance of personal, interpersonal, and personal extra interests. To create a balance between adapting to current settings, influencing existing environments, and selecting suitable activities in the short and long term.

Fourth, the article explains how wisdom in particular and tacit knowledge in general are measured. Fifth, it discusses how education could promote wisdom. Sixth, the article describes active work aimed at helping high school students develop wisdom. The article goes on to suggest that schools can benefit from emphasising the development of wisdom [11]. The researcher found through research and the professional development activities of leading lecturers and lecturers that highly educated professionals often do not know what would be best for them in terms of personal development. In addition, they seek reliable information about what pedagogy they can use to support and enable the development of students' personal wisdom [12]. The author believes that these skills include the capacity to plan and conduct research, gather, organise, analyse, and synthesise information, apply knowledge in novel contexts, and monitor and enhance learning and performance, group work, successful communication in a variety of circumstances. The aim of the curriculum should be independent learning [13]. The researcher states that wisdom, defined as the capacity to respond to actual or possible challenges with options that optimise the prosperity of all involved, both now and in the future, brings tremendous and profound benefits not only to those purportedly unusual individuals who possess it as well as those whom it provides empirically proven social, economic, psychological, and even physical health benefits, additionally to countless other people affected by a wise decision [14]. Proper research on the realization of student wisdom in adding knowledge is the need of the hour for the implementation of NEP 2023.

C. Relevance of knowledge-based education: Need for implementing NEP 2020

Although present education reflects the prevailing culture, ideology and philosophy, it does not necessarily serve the interests of all pupils. The educational system is founded on Western educational theory and philosophy, which are the dominating fields of pedagogy. Western educational theory and philosophy, as a field of dominant pedagogy, constitute the foundation of the educational system, but while it represents dominant cultural ideology and philosophy, it does not necessarily suit the needs of all pupils [15]. Creating a pedagogical approach. Ta's philosophy provides an indigenous knowledge-based pedagogical approach to teaching that has the ability to not only revolutionise learning for all students, but also to equip educators with a comprehensive and interrelated notion of optimal practice. This article investigates how experienced social science research techniques teachers who are experts in their area (method) instruct others on advanced social research methods. The idea of pedagogical content knowledge (PCK), which is well-known in teacher education, is being employed in research method training for the first time. One important component is the teacher's teaching skills, especially pedagogical understanding. Pedagogical knowledge is ongoing knowledge related to teaching and learning, such as understanding learning theory, classroom management, and student motivation. We investigated the functional pedagogical knowledge of 77 teachers who adopted active learning teaching by analysing video clips of large active learning biology courses. To find out and characterise the pedagogical knowledge used by teachers, we used qualitative content analysis and cognitive and sociocultural approaches to learning. We use those ideas shared by teachers to create a pedagogical knowledge framework for teaching active learning in large introductory biology courses [16]. Equivalence and relevance based on time frames: When we try to plot the relevance and equivalence of skills, knowledge and wisdom in differences such as pre-British, medieval and modern, it looks like this (Figure 2.1).

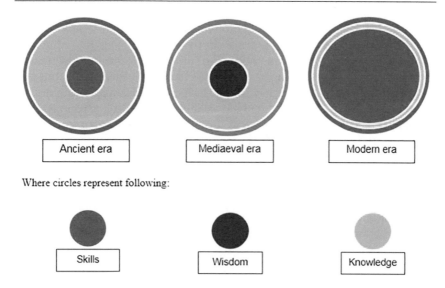

Figure 2.1 Relevance with reference to time frame.

2.4 EVOLUTION OF PEDAGOGICAL APPROACHES WITH RESPECT TO SKILLS, KNOWLEDGE AND WISDOM

Ancient era: In ancient India, education was deeply rooted in the guru-shishya (teacher-student) tradition, which emphasised a personalized and holistic approach to learning. The guru, or teacher, held a revered position and imparted knowledge to their disciples through oral instruction and experiential learning. This pedagogy focused on values, character development, and practical skills relevant to everyday life. The gurukul system was prevalent during this period, where students resided with their gurus and imbibed knowledge not only through formal instruction but also by observing and emulating their teachers' behaviour. Education encompassed a wide range of subjects, including philosophy, mathematics, astronomy, medicine, music, and more. Learning was primarily conducted in small groups, fostering individual attention and mentorship. In this era, skills were majorly focused on life-saving skills. Due to the limitations of the access to education and techniques of teaching-learning, skills majorly remained limited to the survival. The role of knowledge was more important in this era as it was trying to bridge the big gap between skills and the wisdom. Skill was important for individual building whereas wisdom was important for nation building and knowledge acted as a reservoir of information and transfer of information.

Medieval era: This period in India witnessed significant changes in teaching pedagogy due to the influence of various rulers. In this era, pedagogy witnessed structural, philosophical and psychological transformation. To rule a nation having strong foundations of knowledge and wisdom was a challenging task but this was achieved through changing the education system by making changes in the pedagogy. The medieval era witnessed significant changes in teaching pedagogy due to the influence of religious and socio-political factors. Education during this period was often dominated by religious institutions and focused on religious texts and teachings. The transmission of knowledge primarily occurred through monastic schools, madrasas, and cathedral schools. Teaching pedagogy in the medieval era was characterised by rote memorization and recitation of texts. Students were expected to reproduce knowledge without critical analysis or personal interpretation. The teacher played a central role as the authority figure, delivering lectures and transmitting knowledge in a didactic manner. Interaction and engagement were limited, with students passively receiving information. This era in education was characterised by a focus on rote learning and memorization. Students were expected to learn facts and figures without understanding the underlying concepts. The curriculum was also very rigid, with little room for creativity or innovation. Instruction was often delivered in a lecture format, limiting student engagement and critical thinking.

This was labor or clerks producing pedagogy. Do what is directed. People were trained with various skills which were required for running the operations of the dynasties. The access to wisdom was kept limited so that citizens did not feel underprivileged and could be ruled for a longer time without any internal opposition to the ruler. In this era knowledge was used as selective transmission of information. Limited access to wisdom and selective but excessive skill development was achieved through pedagogical changes.

Modern India: The advent of the modern era brought significant changes to teaching pedagogy, driven by scientific advancements, industrialization, and the spread of formal education systems. The emergence of standardized curricula, examinations, and institutionalization marked a shift towards a more structured and uniform approach to education. In the modern era, teaching pedagogy has become more formalized and centralized, with educational institutions playing a crucial role in knowledge dissemination. Teachers adopted a more didactic approach, relying on textbooks, lectures, and examinations to convey information. The focus shifted towards subjects such as science, mathematics, and literature, neglecting traditional knowledge systems. The use of English as a medium of education grew widespread, resulting in a reduction in the use of indigenous languages. Efforts have recently been made to encourage a more student-centred and holistic approach to teaching and learning. Progressive pedagogical methods, such as experiential learning, collaborative activities, and project-based learning, are gaining popularity. Technology has also played a significant role in transforming teaching pedagogy, enabling access to a vast range of educational resources and facilitating interactive and personalized learning experiences. In this era, with the advent of technology, democratization of knowledge is achieved. Knowledge came on finger tips which enabled broadening of wisdom and required skill sets. This era is also being witnessed as an era of entrepreneurship, as a result the role of knowledge became more outcome based rather than output based.

2.4.1 Objective 3: The third objective of the research implies the implementation of "Skills – Knowledge – Wisdom" for NEP 2023

Skills – How to do? Skills are the ability to do something. They are learned through practice and experience. Skills can be physical, such as the ability to play a sport or the ability to type. Mental abilities, such as the ability to solve arithmetic problems or compose a compelling essay, are examples of skills. Skills are important for success in life. They allow us to do the things we want to do and to achieve our goals.

Knowledge – What to do? Knowledge is the understanding of facts or information. It is gained through education, experience, and observation. Knowledge can be factual, such as the knowledge of the capital of France. Knowledge can also be conceptual, such as the knowledge of how to solve a math problem. Knowledge is important for understanding the world around us and for making informed decisions.

Wisdom – Why to do? Wisdom is the ability to use knowledge and experience to make sound judgments and decisions. Wisdom is gained through life experiences, reflection, and learning from mistakes. Wisdom is often characterised by the ability to see the big picture, to understand the consequences of one's actions, and to make decisions that are in the best interests of oneself and others. Wisdom is a valuable asset in life. It allows us to live a more fulfilling and meaningful life. There is correspondence and relevance between Skills, Knowledge, and Wisdom.

Skills, knowledge, and wisdom are all important for success in life. They work together to help us achieve our goals. Skills allow us to do the things we need to do to achieve our goals. Knowledge provides us with the information we need to make informed decisions. Wisdom helps us to use our knowledge and skills to make sound judgments and decisions. The more skills, knowledge, and wisdom we have, the better equipped we are to succeed in life.

2.5 CONCLUSION

The evolution of teaching pedagogy in India has been shaped by cultural, historical, and colonial influences. From the guru-shishya tradition to the formalized British system and the contemporary focus on student-centred approaches, Indian education has undergone significant transformations. As education progresses into the 21st century, there is a growing recognition of the need for student-centred pedagogies that promote critical thinking, creativity, and practical skills. While recognizing the value of modern pedagogical methods, it is crucial to integrate traditional wisdom and indigenous knowledge systems to create a well-rounded and inclusive education system. India can pioneer the way for a genuinely transformational and comprehensive education experience by implementing a balanced approach that respects both cultural history and the needs of the modern world. According to [6] in the book Reimagining Indian Universities, the rigid curriculum, wrote learning, low dedication and competency on the part of the teachers, and weak structural supports for student mobility; almost no research culture, little or inadequate research, a strict and defective evaluation system, poor teaching and learning approaches, and inadequate levels of skill development among students leading to low employability are all problems; overpowering private interests. The quality of higher education and research in India has suffered as a result of these issues and a lack of accountability, despite the reality that the issue is more with ineffective management than a lack of resources. The work further expands the research area of the mentioned areas for effective implementation of NEP 2023.

REFERENCES

[1] https://timesofindia.indiatimes.com/readersblog/educational-blog/implementation-of-new-education-policy-in-india-an-insight-48262/.
[2] R. B. A. A. S. Mangesh M. Ghonge, "Indian Education: Ancient, Medieval and Modern," 2023.
[3] Gupta, https://www.linkedin.com/pulse/transformation-indian-education-system-from-ancient-period-gupta/.
[4] http://www.vkmaheshwari.com/WP/?p=512.
[5] https://www.sociologygroup.com/indian-education-system-features-pros-cons/.
[6] S. R. D. P. Pankaj Mittal, "Reimagining Indian Universities," Association of Indian Universities, AIU House, 16 Comrade, New Delhi, 2020.
[7] Balaram, S., "Design Pedagogy in India: A Perspective," Design Issues, vol. 21, no. 4, pp. 11–22. 2005. http://www.jstor.org/stable/25224015.
[8] Gupta, "The Role of Pedagogy in Developing Life Skills," *Margin: The Journal of Applied Economic Research*, vol. 15, no. 1, pp. 50–72, Feb. 2021, doi: 10.1177/0973801020974786.
[9] Thomas, "Critical Thinking, Transformative Learning, Sustainable Education, and Problem-Based Learning in Universities," *Journal of Transformative Education*, vol. 7, no. 3, pp. 245–264, Jul. 2009, doi: 10.1177/1541344610385753.
[10] R. J. Sternberg, "Why Schools Should Teach for Wisdom: The Balance Theory of Wisdom in Educational Settings," *Educational Psychologist*, vol. 36, no. 4, pp. 227–245, Dec. 2001, doi: 10.1207/s15326985ep3604_2.
[11] Z. M. Rhea, "Old Pedagogies for Wise Education: A Janussian Reflection on Universities," *Philosophies*, vol. 6, no. 3, p. 64, Aug. 2021, doi: 10.3390/philosophies6030064.

[12] W. M. W. Yusoff, R. Hashim, M. Khalid, S. Hussien, and R. Kamalludeen, "The Impact of Hikmah (Wisdom) Pedagogy on 21st Century Skills of Selected Primary and Secondary School Students in Gombak District Selangor Malaysia," *Journal of Education and Learning*, vol. 7, no. 6, p. 100, Sep. 2018, doi: 10.5539/jel.v7n6p100.

[13] M. Bracher, "Foundations of a Wisdom-Cultivating Pedagogy: Developing Systems Thinking across the University Disciplines," *Philosophies*, vol. 6, no. 3, p. 73, Sep. 2021, doi: 10.3390/philosophies6030073.

[14] H. Forsyth, "Āta: An Indigenous Knowledge Based Pedagogical Approach to Teaching," *Universal Journal of Educational Research*, vol. 5, no. 10, pp. 1729–1735, Oct. 2017, doi: 10.13189/ujer.2017.051009.

[15] M. Nind, "A New Application for the Concept of Pedagogical Content Knowledge: Teaching Advanced Social Science Research Methods," *Oxford Review of Education*, vol. 46, no. 2, pp. 185–201, Mar. 2020, doi: 10.1080/03054985.2019.1644996.

[16] A. K. Auerbach and T. C. Andrews, "Pedagogical Knowledge for Active-Learning Instruction in Large Undergraduate Biology Courses: A Large-Scale Qualitative Investigation of Instructor Thinking," *International Journal of STEM Education*, vol. 5, no. 1, p. 19, Apr. 2018, doi: 10.1186/s40594-018-0112-9.

Chapter 3

Content and usability of MOOC platforms for e-learning
An evaluation in higher education

Adarsh Garg[1], P Pradeep Kumar[2], and Ravinder Rena[3]
[1]GL Bajaj Institute of Management and Research, Greater Noida, India
[2]Galgotias University, Greater Noida, India
[3]Faculty of Management Sciences, Durban University of Technology (DUT), Durban, South Africa

3.1 INTRODUCTION

The success of any country depends upon the development of every citizen, which can be achieved through education. The education systems across societies have undergone a paramount change, keeping in line with the changing course of time. This also affects the way of life and how teachers and students interact, their role in society, etc. From classroom teaching, the education sector has slowly moved towards online platforms, which provide a more advantageous opportunity for students from remote areas or suffering from disease, etc [1].

An online learning platform consists of a webspace or portal which is dedicated to educational content and resources. Such platforms offer a student the required educational resources such as lectures, notes and all other essential things in a single place. It also provides opportunities to meet and chat with other students and is a brilliant method for the student and the teacher to interact from a distance and monitor progress effectively.

3.1.1 Types of online learning platforms

Online platforms underwent numerous variations during the rapid advancements made in learning through technology. Such changes could be arranged into different groups according to their geographical description, scope of learning, research approach, and methodology used. These features make division of the online learning platform diverse. Some of the well-known types of online learning platforms can be divided into collaborative e-learning, lifelong learning, m-learning, and personalised e-learning [2].

1. **Collaborative e-Learning:** This type of platform provides an interactive interface so that learners can interact with one another, leading to swift and efficient exchange of knowledge.
2. **Lifelong Learning:** Lifelong learning refers to the traditional method of learning which takes place throughout life. It can be informal, formal, or natural. The primary aim is to enhance the knowledge base and skills of a person residing in a society or for employment purposes.
3. **m-Learning:** The use of mobile technologies and the internet has been exploited in this type of learning platform. They provide free educational materials to learners anywhere and anytime.
4. **Personalised e-Learning:** Personalised online learning systems develop educational materials based entirely on learner's requirements. Such personalised learning is destined to be the next-generation medium of education.

Depending upon the way in which courses are presented, online learning platforms can be broadly divided into the following five types [2]:

1. **Learning Destination Sites:** These types of sites provide online learning courses from various providers in a shared platform. Learners can browse through the content and take up any course in which they are interested. These courses can be free or availed at a certain price. An authoring tool or a learning management system is usually employed in a learning destination site. Typical examples of learning destination sites include Coursera, Udacity and edX.
2. **Traditional Commercial Learning Management Systems:** These types of platforms are more concerned with corporate learning and development and these platforms provide fundamental skills necessary for designing and presenting an online course. The products are usually developed for training within a company. These types of platforms equip course designers with profiles to store and manage their learning systems and these have grading systems to monitor progress.
3. **Modern Learning Management Solutions:** The main disadvantage of traditional learning management solutions is that the courses are not completely tailored to the clients' needs. This is remedied in modern learning management solutions, where companies focus more on learners' needs. Typical examples of modern learning management solutions include Docebo, Motivis, Totara, etc. However, these learning systems are comparatively costly.
4. **Learning Management Ecosystems:** In these types of platforms, all necessary learning engines, integrated tools, e-commerce sites and course authoring are combined and presented as a single solution. These sites usually have a custom front end. An example is the University of Notre Dame's NeXus platform, which uses Open edX

(an open source Learning Management System (LMS)). The system presents a collection of all software required in a single platform.
5. **Custom Built Platforms:** These are customised solutions made from scratch, considering the business server and the learner. A relevant example is the HBX, a customised platform built for Harvard Business School that provides high-end business-related certified courses.

Due to fast changing technological resources and advancements, online or e-learning has undergone several changes. As such, these learning platforms can be defined and divided based on their scope of learning and research, geography used for this purpose as well as the methodology employed. Such approaches have made classification of these platforms varied. One such platform is the MOOC platform. The concept of MOOC can be linked to the introduction of distance learning in Great Britain over 150 years ago [5]. The main purpose was to provide quality education to people who had geographical disadvantages or faced financial crisis to attend regular institutions. Originally, the exchange of knowledge was strictly between the sender and the receiver, with no or minimum interaction. However, with the advent of computers, these exchanges converted into quality interaction between learners and their tutors, or with other students. This method of distance learning further inspired the formation of numerous learners on the internet, giving rise to Massive Open Online Courses (MOOCs) [3–5].

MOOCs were approximately introduced in 2008 and described courses with their underlying connectivity. Compared to previous years, connectivism is considered to be significant in the modern age. Till 2012, MOOC was restricted to only academic usage and did not receive much attention in terms of commercialisation. However, from 2012, with the involvement of top tier institutes such as MIT, Stanford and Harvard, MOOC received the much-needed appreciation and spotlight. MOOCs grew in popularity with higher funding and academic leaders. Platforms such as Udacity, Coursera and edX received recognition. Since then, MOOC has been a top platform for imparting knowledge, enabling self-learning as well as successful interactive sessions between concerned individuals [1].

3.1.2 MOOC

The term MOOC was first coined by Dave Cormier in 2008. The MOOC platform chiefly defines a site for a flexible and effective way to enhance skills and take courses for advancement of careers at a reasonable cost. Due to its non-rigid features, acquiring knowledge at an MOOC platform provides learners with ample opportunities—they can learn at their own suitability and time and can choose what and how to learn [6].

MOOCs allow learning through audio-visual, visual and audio, and the learner can choose any one depending upon the preference. The primary goal of MOOCs was to provide easily accessible and fundamental knowledge to all [1]. The main characteristics of MOOCs can be defined as —openness, persistence constraints and models. Openness refers to the open-source availability of information flow, educational resources, related technological information, organisational overview and other related knowledge. These resources are made available to all participants. The persistence of the learners in MOOC mode faces several barriers compared to life-long learning system. Numerous dropouts in MOOCs can be observed when the learners are unable to keep up with the time issue or when new trends in technology come up. All these issues are some distinct properties of MOOCs. One of the beneficial properties of MOOCs is that they offer multiple learning models, in terms of content delivery, assessment, approach and mode of presentation. This property gives MOOC platforms a bigger advantage compared to traditional learning methods [3].

MOOCs can be broadly divided into the following two types—cMOOCs and xMOOCs [7,8]. cMOOCs mode depends on user participation and originality by emphasising a wide network of online connections. It is based on the ideologies of connectivity, openness and shared teaching. In cMOOC, the massive learning platform is developed on the concept of shared knowledge through interactive sessions and collective contributions. Therefore, the typical participants in cMOOC are self-organised with regard to their knowledge base, learning goals and shared interests. xMOOCs are more concentrated on transmission of university level knowledge across a wide number of students. It is based on a more cognitivist-behaviourist model. xMOOCs consist of numerous video and audio presentations, texts and several automatic assessment methods. xMOOCs follow a method which is more inclined towards the traditional method of teaching.

3.1.2.1 Adoption of MOOC in higher education

Recently, the MOOC platform has attracted a substantial number of persons related to higher education being an open-access, massive and online platform for learning [9,10]. For the quick adoption of this new technology, MOOCs have been organised to provide quality educational resources, solutions to their problems and chances to get the benefit of world class education staying in their native place where limited opportunity for traditional education persists [6,11,12]. In contrast to the expected results, the adoption rate is not showing positive results [13,14]. A survey conducted in February 2013 on the professors taking MOOC classes suggested that the number of enrolments per class was quite attractive (almost 33,000 on average per class; Chronicle of Higher Education). However, the passing rate was quite low (on average 7.5%).

The work in [15] tried to compile information about enrolment and completion of 91 different types of MOOCS. The study showed a median enrolment figure of approximately 42,000, but the number of enrolments decreased over a period. The association between course length and enrolment numbers has been studied also and it showed that the longer the courses, the more enrolment of students it attracts. Completion of a course was evaluated by students obtaining course certificates. For 39 courses, the study showed that the completion rate varied from 0.9% to 36.1% with a 6.5% median value. Such low rate of completion was maintained among all the students from different universities and different seasons. However, course length showed a substantial effect on the rate of completion. The longer the course, the lower the percentage of students who finished the course.

The work in [16,17] studied the registration and course completion data on MOOC provided through the platform of edX by MIT and Harvard University. They found out some key facts regarding the learners of this online educational platform. As registration in MOOCs does not require any cost or commitment, the traditional measure of a successful course may not be able to describe the course engagement by the learners. For example, a skilled learner can visit to access any specific content of the course and he/she may not need the certification. Therefore, the statistics of enrolment and registration or course completing participants may not fully evaluate the course success. Additionally, the work in [16] showed that the acceptance of MOOCs is geographically biased to a major extent. The top 25 countries in order of the number of registrants for MOOCs were assessed by locating their IP address or email address during registration. It showed that very few to no registrants are from developing countries, and a majority of the participants are from the United States of America.

The work in [14] highlighted the patterns emerging from the previous dataset showing poor adoption rate of MOOC among students. The three obvious patterns are: (1) MOOC learners are leaving the course after one year and never returning; (2) Participants from developing countries are not increasing in a significant manner; and (3) the lowered rate of course completion by the learners are not improving since the initiation of MOOC.

Hence, it is essential to comprehend the different possible elements that enhance the adoption of MOOCs.

3.1.2.2 Factors affecting MOOC

Research illustrates that the use of MOOC has gained much attention, especially among working professionals, who are aiming for either a higher education or to enhance their skills associated with their jobs [18]. The MOOC platform has grown over ten-fold in recent years compared to the traditional methods of education and has helped numerous individuals. Most of the studies concerning MOOCs platforms have been concentrated on handling and finding technical issues as shown in Table 3.1. However, limited studies have addressed the issues of content quality and usability issues of MOOCs platforms. Thus, this study puts forward an evaluation of content and usability of MOOCs platforms with reference to existing studies.

Content and usability of MOOC platforms for e-learning 33

Table 3.1 Factors influencing adoption of MOOCs

Author and year	Significant factors	MOOC platform/s	Location
Meet et al. (2022) [19]	Effort expectancy; Performance expectancy; Facilitating condition; Hedonic motivation; Price value	SWAYAM	India
Chaveesuk et al. (2022) [20]	Habit and Price value; Social influence; Hedonic motivation; Facilitating condition; Culture; Effort expectancy	Navoica, Coursera edX, SkillLane, MUx, KMITL Learning Intelligence X, and CHULA MOOC	Poland, Thailand and Pakistan
Gupta and Maurya (2022) [21]	User characteristics, technological characteristics, and features of online courses	Not mentioned	Delhi, India
Khalid et al. (2021) [22]	Absorptive capacity; Social influence; Perceived autonomy; Facilitating conditions	Any MOOC	Thailand and Pakistan
Wan et al. (2020) [23]	User satisfaction; Social influence; Effort expectancy; Performance expectancy	ICourse and Netease Opencourse	China
Pozón-López et al. (2020) [24]	Autonomous motivation; Perceived satisfaction; Perceived ease of use	Abierta UGR, Crehana, EducaLAB and Tutellus	Spain
Aparicio et al. (2019) [25]	Information quality, Service Quality, System quality; Gamification	xMOOC	Various country
Mulik et al. (2018) [26]	Perceived value; Facilitating conditions; Social influence; Effort expectancy; Performance expectancy	Any MOOC	Any location
Khan et al. (2018) [27]	Perceived competence; Perceived relatedness; Social recognition; Task and technology characteristics	Any MOOC	Pakistan

(Continued)

Table 3.1 (Continued) Factors influencing adoption of MOOCs

Author and year	Significant factors	MOOC platform/s	Location
Fianu et al. (2018) [28]	Performance expectancy; Computer self-efficacy; System quality; Facilitating conditions; instructional quality; MOOC usage intention	Any MOOC	Accra, Ghana
Zhang et al. (2017) [29]	E-learning self-efficacy; Personal innovativeness; Learner control; Perceived ease of use; Perceived usefulness	Coursera and ICourse163	China
Barak et al. (2016) [30]	Intrinsic motivation and self-determination	Coursera	Various countries
Alraimi et al. (2015) [31]	Perceived and user satisfaction; Perceived usefulness; Perceived openness; Perceived reputation	Coursera, edX and Udacity	Various countries

Source: Author's own after literature review.

3.1.2.3 Content of MOOCs

Due to the high-quality content provided in the MOOC platform, it can be assumed as being an alternative to traditional methods of learning [19]. The content is provided under different formats, which can include audio, visual or audio-visual presentations, depending upon the objective. Some of the contents are provided in a pre-defined format which is designed by the course instructor. Some of the contents are flexible and give the learners room to create the content according to their time, availability and level of understanding. Additionally, the MOOC platform also provides interactive platforms for user and creator interactions as part of the content [20]. Such properties of MOOC content make the platform user friendly and innovative.

Based on literature, similarities among the different content in MOOC can be observed. The most significant is that all the content has been defined with alignment of the objective and it has been delivered by a team of instructors or educators. The content typically entails the information of the course and assessment of the learners understanding. The primary people involved in creation of this content are—educators, teaching assistants, staff members and mentors. The educators or instructors were assigned the duty of overseeing the entire course from designing to learners' involvements in the course. Any technical issues along with helping the educators were provided by the teaching assistants and the staff members. The mentors typically provided voluntary services with prior experience of joining MOOC courses. The mentors are public figures so that the present learners can voice their doubts directly [3].

The content of MOOCs must serve the purpose of improvement of a learner's life through knowledge. Open course content provides such opportunities to all individuals, even with distance issues [2]. Such a content design helps in eradicating the knowledge scarcity prevalent among citizens, especially those of developing countries such as India [21].

The following are some of the significant attributes to be considered for content in an MOOC platform:

- Widely sought-after content must be made easily and freely available
- The content must be curated to the needs of the learners
- The skills and capabilities of the target audience must be taken into consideration while designing any content
- Experts in specific domains must be involved to make content more relevant and efficient

Relevant, updated content with rich collaborative information is very important for a MOOC to be effective and successful. It is pertinent to acknowledge that MOOC comprises activities, communication, and actions, and it cannot be rendered as free online content. Various interconnected nodes of knowledge facilitate learning. Due to the rapid changes in content, it is less important than the pipe in which it is provided [6]. The content of MOOCs comprises discussion forums, projects, multiple-choice questions, short quizzes, and resource-related short movies [22]. The success of MOOCs is determined by the quality of course content, which is one of the crucial components. The basic features of a quality course content are evaluated on the basis of 1) MOOC design, 2) the relevance of the course; 3) The ease of understanding the course materials, and 4) updated version of the material. Albelbisi (2020) studied different factors leading to MOOC success and course quality was one of the vital factors. Course quality was measured by five questionnaire items [23].

The work in [24] introduced a gamified model (Figure 3.1) of MOOCs in Universidad Rey Juan Carlos in Spain. Gamified MOOCS increased engagement and participation of learners by introducing game elements such as completing a task in a scheduled time, discussing on any topic with peers and posting comments on that topic, and receiving reward points or promotion over their engagement to the courses. The findings of the study demonstrated that apart from variables of D&M model (system quality, service quality, information quality), gamification had a significant effect on the use of MOOC. Therefore, recent MOOC contents are adopting gamification in significant manners.

Rajas-Fernández et al. (2021) showed the potential of audio-visual medium in the content making. They analysed the content of the videos,

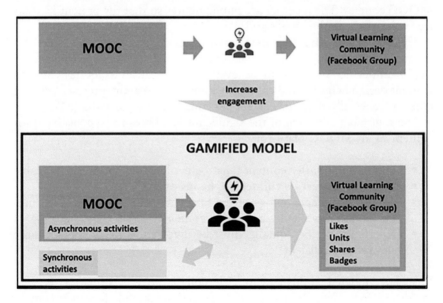

Figure 3.1 Gamified model as conceptualised.
Source: Borrás-Gené et al. (2019).

the audio-visual languages used in these videos, and common characteristics of such material. In the analysis, they targeted introductory videos for MOOC courses from the Miríadax platform. This study proposed several genres of video lectures such as fully animated, partly animated, semi-live action and semi-animated video, mainly live action video. Apart from that, addition of cultural aspects in the content showed to be helpful. The content has been made in story-telling perspective and the storytellers are the instructors in a cartoon form with similar cultural appearance as the learners. Research has shown learners are attracted more towards the same race or culture due to facial attractiveness as compared to strangers from a different cultural background. Communication via video lectures is also an important aspect for the MOOC learners. Majorly in developed MOOCs, informal communication styles are used rather than a formal one. This is because an informal way of approach is beneficial for learning. The questionnaire-based study showed that most users were gratified with the animated course content.

3.1.2.4 Usability

Usability of MOOC platform refers to the effective achievement of imparting knowledge to the concerned students in the online platform; without using the traditional educational tools [7]. Several parameters have been illustrated in the literature to measure the usability of MOOC. In this regard, MOOC platform has a high degree of usability if it provides interactive tools to aid self-design of courses while maintaining a relatively similar outlay reducing any cognitive load [25]; usability is defined as the rate of satisfaction of a learner while studying a course. Besides cognition and satisfaction, the work in [26] stated that usability can be increased by incorporating problem solving skills into the courses. However, the work in [27] highlights that all these various factors such as content, familiarity of navigation, interactivity, assessment and support can affect the motivation of the learners; which in turn defines the usability.

While evaluating usability, issues such as dissatisfaction with courses and perceiving powers of the learners, need to be considered. The work in [7] states that an approach and feedback checklist must be incorporated to evaluate the usability of a course. Table 3.2 displays the significant factors to be considered while evaluating usability [6].

3.2 DISCUSSION AND CONCLUSION

Students across the world use online courses to improve their skills and knowledge. As such, an MOOC provides an efficient platform for learners to have access to a wide variety of effective study material. Therefore, research in the direction of the content and its usability of courses is

Table 3.2 Usability traits

Satisfaction	User feedback on how satisfied they are after attending the course
Efficiency	Degree of productivity of the course
Cognitive load	Reduce it by employing familiar layout
Interactions	Interactive forums to clear doubts, etc.
Learnability	Courses are easy to comprehend
Error	Users can recover easily from errors while navigation
Effective	Courses are effective and achieve target goals of users
Visually stimulating	Enjoyable course structure through interesting visual inputs
Feedback	A proper feedback structure for assessments

Source: Author's own.

imperative. The overall analysis presented in this study suggests that content and usability are associated and must be considered to significant factors while designing any course in an MOOC platform. The content must be designed to encompass certain factors such as easy availability, useful information and designed keeping in mind the target audience. Further, the content must be such that it can be easily understandable. The degree of usability of such content may be evaluated by considering several factors. Among them, the satisfaction of learners is of utmost importance. Such assessment can be made by employing effective feedback tools. For other aspects, easy to use with less application of cognitive load, easy navigation, flexibility and simplicity are some of the factors to be considered to increase usability of content. The critical analysis thus finds significance in considering content and usability for the success of any course available in the MOOC platform.

REFERENCES

[1] Ospina-Delgado, J. E., Zorio-Grima, A., & García-Benau, M. A. (2016). Massive open online courses in higher education: A data analysis of the MOOC supply. *Intangible Capital*, 12(5), 1401–1450. 10.3926/ic.798

[2] Zammit, M. L., Spiteri, J., & Grima, S. (2018). *The Development of the Maltese Insurance Industry: A Comprehensive Study*. Emerald Group Publishing.

[3] Mohamed, M. H., & Hammond, M. (2018). MOOCs: A differentiation by pedagogy, content and assessment. *The International Journal of Information and Learning Technology*, 35(1), 2–11. 10.1108/IJILT-07-2017-0062

[4] Mosharraf, M., & Taghiyareh, F. (2016). The role of open educational resources in the eLearning movement. *Knowledge Management & E-Learning: An International Journal*, 8(1), 10–21.

[5] Ng'ambi, D., & Bozalek, V. (2015). Massive open online courses (MOOCs): Disrupting teaching and learning practices in higher education. 10.1111/bjet.12281

[6] Zhou, Q. (2018). *Usability Study of Massive Open Online Courses (MOOCs) Platforms [Master of Science]*. Northeastern University.

[7] Syahid, A., Kamri, K. A., & Azizan, S. N. (2021). EBSCOhost l 153144080 l Usability of massive open online courses (MOOCS): Malaysian undergraduates' perspective. 18(3). 10.9743/jeo.2021.18.3.11

[8] Totschnig, M., Willems, C., & Meinel, C. (2013, May). openHPI: Evolution of a MOOC platform from LMS to SOA. In *International Conference on Computer Supported Education* (Vol. 2, pp. 593–598). SciTePress. 10.5220/0004416905930598

[9] Alraimi, K. M., Zo, H., & Ciganek, A. P. (2015). Understanding the MOOCs continuance: The role of openness and reputation. *Computers & Education*, 80, 28–38. 10.1016/j.compedu.2014.08.006

[10] Veletsianos, G., & Shepherdson, P. (2016). A systematic analysis and synthesis of the empirical MOOC literature published in 2013–2015. *International Review of Research in Open and Distributed Learning*, 17(2), 198–221. 10.19173/irrodl.v17i2.2448.

[11] Garg, A. (2021). Online education- A learner's perspective during COVID-19. *Asia Pacific Journal of Research and Innovations*, 16(4), 279–286. 10.1177/2319510X211013594.

[12] Garg, A. & Garg, K. (2021). Emerging technologies to sustain students' engagement in learning. *Rabindra Bharati Journal of Philosophy*, XXIII(12), 111–118.

[13] Koutropoulos, A., Gallagher, M. S., Abajian, S. C., de Waard, I., Hogue, R. J., Keskin, N. O., & Rodriguez, C. O. (2012). Emotive vocabulary in MOOCs: Context & participant retention. *European Journal of Open, Distance and E-Learning*. Vol. n1, 1–23. https://eric.ed.gov/?id=EJ979609

[14] Reich, J., & Ruipérez-Valiente, J. A. (2019). The MOOC pivot. *Science*, 363(6423), 130–131. 10.1126/science.aav7958

[15] Jordan, K. (2014). Initial trends in enrolment and completion of massive open online courses. *International Review of Research in Open and Distributed Learning*, 15(1), 133–160. 10.19173/irrodl.v15i1.1651

[16] Ho, A., Reich, J., Nesterko, S., Seaton, D., Mullaney, T., Waldo, J., & Chuang, I. (2014). HarvardX and MITx: The first year of open online

courses, Fall 2012-Summer 2013 (SSRN Scholarly Paper No. 2381263). *Social Science Research Network*. HarvardX and MITx Working Paper No. 1. Posted: 22 Jan 2014, 1–33. 10.2139/ssrn.2381263.
[17] Siemens, G. (2006). Knowing knowledge. *Lulu.com*.
[18] Sharma, Y. (2017). Global: Move Over Moocs – Collaborative Mooc 2.0 is Coming. In G. Mihut, P. G. Altbach, & H. de Wit (Eds.), *Understanding Global Higher Education: Insights from Key Global Publications* (pp. 167–169). SensePublishers. 10.1007/978-94-6351-044-8_36
[19] Meet, R. K., Kala, D., & Al-Adwan, A. S. (2022). Exploring factors affecting the adoption of MOOC in Generation Z using extended UTAUT2 model. *Education and Information Technologies, 27*(7), 10261–10283.
[20] Chaveesuk, S., Khalid, B., Bsoul-Kopowska, M., Rostańska, E., & Chaiyasoonthorn, W. (2022). Comparative analysis of variables that influence behavioral intention to use MOOCs. *Plos one, 17*(4), e0262037.
[21] Gupta, K. P., & Maurya, H. (2022). Adoption, completion and continuance of MOOCs: A longitudinal study of students' behavioural intentions. *Behaviour & Information Technology, 41*(3), 611–628.
[22] Khalid, B., Chaveesuk, S., & Chaiyasoonthorn, W. (2021). Moocs adoption in higher education: A management perspective. *Polish Journal of Management Studies, 23*.
[23] Wan, L., Xie, S., & Shu, A. (2020). Toward an understanding of university students' continued intention to use MOOCs: When UTAUT model meets TTF model. *Sage Open, 10*(3), 2158244020941858.
[24] Pozón-López, I., Higueras-Castillo, E., Muñoz-Leiva, F., & Liébana-Cabanillas, F. J. (2021). Perceived user satisfaction and intention to use massive open online courses (MOOCs). *Journal of Computing in Higher Education, 33*, 85–120.
[25] Aparicio, M., Oliveira, T., Bacao, F., & Painho, M. (2019). Gamification: A key determinant of massive open online course (MOOC) success. *Information & Management, 56*(1), 39–54.
[26] Mulik, S., Srivastava, M., Yajnik, N., & Taras, V. (2020). Antecedents and outcomes of flow experience of MOOC users. *Journal of International Education in Business, 13*(1), 1–19.
[27] Khan, A. U., Khan, K. U., Atlas, F., Akhtar, S., & Farhan, K. H. A. N. (2021). Critical factors influencing MOOCs retention: The mediating role of information technology. *Turkish Online Journal of Distance Education, 22*(4), 82–101.
[28] Fianu, E., Blewett, C., Ampong, G. O. A., & Ofori, K. S. (2018). Factors affecting MOOC usage by students in selected Ghanaian universities. *Education Sciences, 8*(2), 70.
[29] Zhang, X., Han, X., Dang, Y., Meng, F., Guo, X., & Lin, J. (2017). User acceptance of mobile health services from users' perspectives: The role of self-efficacy and response-efficacy in technology acceptance. *Informatics for Health and Social Care, 42*(2), 194–206.

[30] Barak, M., Watted, A., & Haick, H. (2016). Motivation to learn in massive open online courses: Examining aspects of language and social engagement. *Computers & Education*, *94*, 49–60.
[31] Alraimi, K. M., Zo, H., & Ciganek, A. P. (2015). Understanding the MOOCs continuance: The role of openness and reputation. *Computers & Education*, *80*, 28–38.

Chapter 4

Machine learning in medical imaging
A comprehensive study

Debanjana Ghosh and Srilekha Mukherjee

University of Engineering & Management, Kolkata, India

4.1 INTRODUCTION

Machine learning (ML) is the field of creating models which can be fed with data to create various analyses or predictions. It has numerous applications in several fields like Finance, Healthcare, Weather, E-commerce, Image Processing, etc. In this report the applications in the domain the medical sciences will be explored. Healthcare is one of the largest fields which broadly explores the benefits of ML. It is changing the way of treating the patients and bringing differences in their day-to-day life. From creating new life-saving drugs to predicting the life-threatening diseases beforehand, ML is spreading its usefulness everywhere.

As technological advancement is happening in the domain of medical practice, Machine Learning is also taking a lead. Machine Learning and Deep Learning have been implemented in different aspects of medical imaging like image acquisition, novel imaging modalities for several types of images like CT, MRI, multi-X imaging, and multi-modal or multi-planar (PET/MRI, PET/CT) technologies, etc. [1]. As a huge amount of data is generated each and every moment in terms of medical images, it becomes challenging for medical professionals to plan and follow up only by themselves. At the same time, the huge amounts of image data generated in every corner of the world are being archived in huge databases of medical information systems to make them more accessible in research studies. This also generates technical challenges and issues related to privacy which have to be dealt with. Hence, ML has a lot of scope to show its applications in the above field. The well-organized data in medical imaging also provide opportunities for data analysis to bring innovation in different wings such as detection of anomalies for individual patients, called computer-aided or computer-assisted diagnosis, discovery of the features to detect disease early i.e., imaging biomarkers, prediction of treatment outcome beforehand to provide optimal therapy, and phenotype and genotype correlation which is imaging genetics.

Medical Image Analysis is all about getting relevant information from medical images. The involvement of only human technicians to analyse the images is time-consuming and impractical, and also prone to human error. Hence, there is a huge need to design efficient and reliable semi-automated or automated models for the analysis of medical images using machine interpretation in everyday medical practice. Medical Imaging often concerns the calibration of definite attributes of the objects of concern i.e., volume, position, symmetry, size, shape, extent, etc., and the evaluation of anatomical alterations over time like aging, motion of organ, deformation of tissue, and abnormal lesion growth. Moreover, it involves characterization recognition of normal versus abnormal development morphological variation between subjects.

There are several methods [2] for medical image analysis. But most of them consist of some basic steps as follows.

4.2 BASIC STEPS

4.2.1 Image segmentation

Image segmentation is about defining the boundaries or edges. It detects the edges around the object of interest in an image and differentiates which image voxels are of a particular object and which are not. It is a prerequisite to quantifying the geometric properties like volume or shape of the object. In can be done in various ways: (a) delineation of the contour of the object by boundary-wise in one i.e., 2D image segments or multitudinous i.e., 3D image segments, (b) section-wise clustering of the voxels which probably belong to the identical object (maybe into one or multitudinous sectors), (c) voxel-wise assignment of every voxel within the image to a distinct object, tissue class or background. Class markers allocated to a voxel are prospective which results in a fuzzy or soft partition of images. If an image is of limited resolution, then 3D segmentation can lead to a loss of contrast and detail. So, to fill those missing details, interpolation is required. In many cases of clinical practice, to simplify things, approximate 1D or 2D analysis is used in place of precise 3D measurement.

4.2.2 Image registration

The spatial relationship is determined by image registration between different images. It is all about matching the image based on its own content [3]. Images taken at different times like before treatment or after treatment, or from different subjects, or with different modalities often contain information that are complementary in nature which can be fused or analysed together at voxel level for using the images' full resolution. It is used and required to rectify differences which are unknown in positioning of the patient in the scanner. After successful registration, those images are resampled in a general space and get amalgamated for joint analysis. When deformations can be ignored, the registration solution is called and presented as an affine transformation matrix where few parameters are used. In other common cases further complicated conversion is needed.

4.2.3 Image visualization

The particulars which were drawn out from the inputs have to be represented in an optimized way for the support of diagnosis and planning of therapy. For the assessment of constructional associations in and between the objects in 2D multi-planner, 3D medical images, visualizations are not effectively suited. Either volume rendering or surface rendering has to be registered. When a 3D segmented part of the object of interest of the images is rendered under definite conditions of lightening by the assignment of properties of material (which specifies its specular and disseminated light reflection, transmission, scattering, etc.), it is assumed as surface rendering. When the image voxels are rendered directly by the specification of proper transfer functions that is used to assign a colour and opacity to each voxel which depend on their intensity, it is called volume rendering. In case of real-life applications of medical imaging like image-based surgery, more tools need to be incorporated to combine the real-world images with virtual reality.

All these methods cannot be considered as separate smaller problems in medical image analysis. They are connected and treated as a solution which is optimal for a specific problem, which can be reached by appraising jointly all decomposition, visualization and registration [3].

4.3 CHALLENGES

Medical image analysis can get intricate due to several reasons. The main challenge lies within the quality and quantity of the input image which is required for analysis. These data are complex in nature which gives rise to the challenges. Next comes the object of interests and its complexities. In medical image analysis, the objects cannot be assessed with complete accuracy due to the lack of data from outside.

These input images are mostly 3D in nature. Hence, they provide additional information along with additional complexity. Rather than processing them in 2D slice by slice, they should be processed in 3D as it analyses all three dimensions together. As medical images are based on different principles, they get complicated due to the ambiguity caused by the image acquisition process. Often there are applications which involve the extraction of complementary information from images. For these various reasons, dealing with images of different nature like multi-dimensional, multi-subject, multi-modal, multi-parametric, multi-centre, and multi-temporal, raises several issues [4].

In every image there are specific anatomical structures which are called objects of interest. They may be rigid or flexible, and may be normal or pathological. These objects have a complex shape that is hard to describe by mathematical models. Moreover, these objects can show intra-subject as well as inter-subject variability both in intensity and shape. These variabilities in different objects of interest pose challenges.

Medical images are sometimes combined with computational tool generated simulated images to add more flexibility to them. But these simulated images are sometimes unable to help apprehend the full intricacy of the real images and create challenges.

4.4 ROLE OF ML IN MEDICAL IMAGE ANALYSIS

Medical image analysis is near the convergence of Medical Computer Vision, Image and Machine Learning. Machine Learning is aiming to develop approaches for medical image analysis which will be able to create a robust and efficient automated or semi-automated model for accurate analysis [4,5]. This kind of model-based analysis should follow two things: the objective derivation function for assessment about how well the specific model fits to the input, and to select a befitting optimization scheme to find the specifications about the model instance which are optimal and prime fit the input data. These models must be sufficiently flexible to incorporate the variability in images due to several factors.

Nowadays, models of supervised learning are advancing. Chiefly deep learning or more specifically convolutional neural networks (CNN), exhibit significant promise in the domain of medical imaging and computer vision. They have made groundbreaking discoveries in image classification, segmentation, object recognition, etc. [5,6]. The problem statement here is basically a classification task, where the input data are the large set of non-specified local features from within or between images. The neural network has to learn by itself. A highly complex function class is defined by it and large amounts of data are required for stable solution and good generalization.

The following are some of the ML algorithms which are typically used for image analysis.

4.4.1 Supervised machine learning

Here a collection of large number of images are manually labelled first by experts: for example whether they are malignant or otherwise. Hence, the first step is to calculate features which have strong correlation with the indication of malignancy. Generally, there is more than one feature which makes feature vector. The point to be noted is that there has to be appropriate pre-processing for making the features reproducible. Then, feature selection or feature reduction is done to reduce less informative features from the feature vector. Class generally refers to different types of output such as yes or no, malignant or benign. After choosing the features they are used as input to any of the ML algorithms like decision tree, support vector machine, Bayes network, neural nets, etc. [7].

4.4.2 Decision tree

This is one of the simple and easy methods to understand supervised learning algorithm. It consists of a series of decisions. The most important decision is taken first. For a collection of examples, the entropy of all the features has to be calculated first. Then the information gained can be calculated and based on that the feature is selected [8]. The weakness of this model is that there is a chance of overfitting and noise as the behindhand resolutions will be determined extremely by the exact input given.

4.4.3 Support vector machine

Support vector machine (SVM) has two basic concepts, one is the plane or support vector which separates two classes and the other is the challenge of plotting points from their original lower dimensional space to a different or higher dimensional space. As all problems cannot be solved using the concept of 2D linear plane, SVM creates the concept of hyperplane. Some tasks of classification are such that they cannot be linearly separable, hence the mappings used by SVM are incorporated in those cases. SVM has parameters, called hyperparameters which check for the penalty due to misclassification of points. Hyperparameter selection and adjustment is very complex and needs an understanding of both images and machine learning algorithms. These algorithms are used as well for outlier identification and regression [9].

4.4.4 Neural networks

This is a form of machine learning where a bridge is made between the understanding of machine and human brain. In case of imaging or medical imaging, input features which are fed to the network are multiplied by weights. Thus, they pass through the nodes of different layers. The input gives rise to the output depending on the 'activation function'. This process continues till the final output. Adjustment of the weights is critical in neural networks and is done by backpropagation. One of the problems is to find which weight is causing error in the output among many weights. This makes the method difficult [9].

4.4.5 Deep learning

Nowadays, deep learning is getting a lot of attention in medical image analysis. In various world-wide challenges based on image classification and analysis, deep learning is showing superior performance to traditional machine learning algorithms. AlexNet, a deep learning algorithm defeated the field of ImageNet competition by more than 10% in 2012 [10]. One more deep learning method produced another 10% gain over the previous in 2013. Now this challenge is dominated by deep learning algorithms where the performance is better than human performance at above 98%. Thus, deep learning has achieved the spotlight in the domain of imaging.

4.4.5.1 Deep learning layers

There are the following layers in a deep learning model:

4.4.5.1.1 Fully connected layer

A traditional neural network is made up of nodes and connectors consisting of weights and activation functions. These are usually addressed as the fully connected layer that is preferably deployed near the end part of the network.

4.4.5.1.2 Convolution layer

These layers are generally placed on the input side, to help the system learn about the critical features from the input. In case of images as input, more than one convolution layer can be used.

4.4.5.1.3 Pooling layer

Pooling layer follows the convolution layer and combines the output of the adjacent convolution layers into a sole output. "Max Pool" is the most popular pooling function. Here, the highest value for its 'window' is found and then passed to the next adjacent layer, which is most of the time another convolution.

4.4.5.1.4 Activation layer

To introduce non-linearity in the system, an activation layer is placed in it. Previously it was a sigmoidal-shaped function but much simpler functions are introduced as the activation function. Nowadays, rectified linear unit or ReLU is a famous activation function that gives 0 for the negative input and 1 for positive input.

4.4.5.1.5 Output layer

This is the final layer and also one type of activation layer. For different tasks, different types of output layer are chosen such as for regression linear output layer or for classification of SoftMax activation layer. In choosing this layer the importance of the cost function is also to be considered.

4.4.5.1.6 Residual layer

Some extra layers are called residual layers. The output of a layer is compared with the identity layer and forces them to do better. It helps in reduction of layers, as well as reduction of parameters which decreases the chances of overfitting of the training data.

4.4.5.2 Types of deep learning models

Depending on the positioning and selection of different types of layers, there are different formations of deep learning.

4.4.5.2.1 Convolutional neural networks

This architecture is used predominantly for the classification of images. Within the input layers, especially convolution and pooling layers, the low-level attributes are effectively drawn out from the image. The pooling layer reduces the resolution by combining low-level features to high-level features. Thus, the initial layers find edges, lines, points, etc. These layers get combined with the following layers to identify objects. Finally, the fully connected (FC) layer is attached with the weights and activation functions to determine the output. Some examples of convolutional neural networks (CNNs) are AlexNet [10], VGGNet [11], GoogLeNet [12], ResNet [13], ResNeXt [14], and region-based CNN [15].

4.4.5.2.2 U-Net

This is one type of CNN. The main characteristic of this network is at the bottom of this architecture, the input is diminished to the key object. Once the key object is recognised, the conversion to the original resolution is attained with the help of bypass layers by upscaling user pixel data. An important instance of an U-Net is SegNet [16].

4.4.5.2.3 Generalized adversarial networks

A generalized adversarial network (GAN) is generally used to create an image rather than segment or classify. Though medical images should be authentic and generated through scanning the patients, this architecture has some application in radiology. It can be used to create additional medical images for training. It is also used to check the system's robustness.

4.5 CONCLUSION

The performance of CNN depends on many factors like, for example properly adjusted weights and training with huge amounts of data. CNN can currently learn faster and better to beat human experts. But it is very important that a huge amount of good training data should be available.

Machhine Learning has been applied in the medical imaging field for decades. It is now getting more spotlight and success than ever [17]. It is an ever-growing field. In the coming days technicians will be more aware and able to implement the usefulness of ML in day-to-day analysis. They should learn about imaging and ML implementations as compulsory to get the full essence of it. In this way they can bring new ideas and explore the use of new products and applications designed by ML in imaging. They will acknowledge the best application of this in medical practice to improve patients' lives as well as the system.

REFERENCES

[1] Suetens, P. (2017). *Fundamentals of medical imaging*. United Kingdom: Cambridge University Press.

[2] Bankman, I. N. (2000). *Handbook of medical imaging: processing and analysis*. Academic Press.

[3] Maes, F., Robben, D., Vandermeulen, D., & Suetens, P. (2019). The role of medical image computing and machine learning in healthcare. *Artificial intelligence in medical imaging: opportunities, applications and risks*, 9–23.

[4] Alanazi, A. (2022). Using machine learning for healthcare challenges and opportunities. *Informatics in Medicine Unlocked*, 100924.

[5] Suetens, P., Fua, P., & Hanson, A. J. (1992). *Computational strategies for object recognition*. ACM Comput Surv.

[6] Greenspan, H., Van Ginneken, B., & Summers, R. M. (2016). Guest editorial deep learning in medical imaging: Overview and future promise of an exciting new technique. *IEEE transactions on medical imaging*, 35(5), 1153–1159.

[7] Barragán-Montero, A., Javaid, U., Valdés, G., Nguyen, D., Desbordes, P., Macq, B., Willems, S., Vandewinckele, L., Holmström, M., Löfman, F., & Michiels, S. (2021). Artificial intelligence and machine learning for medical imaging: A technology review. *Physica Medica*, 83, 242–256.

[8] Gini, C. (1913). Variabilita e Mutabilita. *J R Stat Soc*, 76, 326.

[9] Erickson, B. J. (2019). Deep learning and machine learning in imaging: Basic principles. *Artificial intelligence in medical imaging: Opportunities, applications and risks*, 39–46.

[10] Krizhevsky, A., Sutskever, I., & Hinton, G. E. (2017). Imagenet classification with deep convolutional neural networks. *Communications of the ACM*, 60(6), 84–90.

[11] Simonyan, K., & Zisserman, A. (2014). Very deep convolutional networks for large-scale image recognition. *arXiv preprint arXiv:1409.1556*.

[12] Szegedy, C., Liu, W., Jia, Y., Sermanet, P., Reed, S., Anguelov, D., Erhan, D., Vanhoucke, V., & Rabinovich, A. (2015). Going deeper with convolutions. In Proceedings of the IEEE conference on computer vision and pattern recognition (pp. 1–9).

[13] He, K., Zhang, X., Ren, S., & Sun, J. (2016). Deep residual learning for image recognition. In Proceedings of the IEEE conference on computer vision and pattern recognition (pp. 770–778).
[14] Xie, S., Girshick, R., Dollár, P., Tu, Z., & He, K. (2017). Aggregated residual transformations for deep neural networks. In Proceedings of the IEEE conference on computer vision and pattern recognition (pp. 1492–1500).
[15] Ren, S., He, K., Girshick, R., & Sun, J. (2015). Faster R-CNN: Towards real-time object detection with region proposal networks. *Advances in neural information processing systems*, 28.
[16] Badrinarayanan, V., Kendall, A., & Cipolla, R. (2017). Segnet: A deep convolutional encoder-decoder architecture for image segmentation. *IEEE transactions on pattern analysis and machine intelligence*, 39(12), 2481–2495.
[17] Litjens, G., Kooi, T., Bejnordi, B. E., Setio, A. A. A., Ciompi, F., Ghafoorian, M., Van Der Laak, J. A., Van Ginneken, B., & Sánchez, C. I. (2017). A survey on deep learning in medical image analysis. *Medical image analysis*, 42, 60–88.

Chapter 5

Platform with anonymity for students to foster in-class participation

Abhishek Deupa[1] and Ruqaiya Khanam[2,3]

[1]Department of Computer Science and Engineering, School of Engineering and Technology, Sharda University, Greater Noida, India
[2]Department of Electronics and Communication Engineering
[3]Center for Artificial Intelligence in Medicine, Imaging & Forensic, Sharda University, Greater Noida, India

5.1 INTRODUCTION

In a large class, when students are asked for responses, students who are not sure about the topic, feel uncomfortable about being selected and students who are shy feel uncomfortable being in the limelight, even if they know the correct response [1]. Students strive to portray a positive image of themselves to their classmates. Due to this, they are often afraid to share information for fear that they will be judged for their mistakes or be the centre of attention by others [2]. Engaging large numbers of students and maintaining their active participation can be challenging, but it helps students better understand the material when they have an active experience [2]. Along with making a positive difference in better understanding topics when doubts were put forward, a better perception of their understanding of lectures and ability to judge their own understanding is significantly higher [1]. Different ways to anonymously elicit responses from students could be by having a show of hands to statements like 'Raise your hands if you think this is the answer' or by distributing coloured papers which represent particular answers, but these methods could be redundantly time consuming and also not totally anonymous [1]. It is the anonymity factor that promotes increased understanding since it induces students to pick a particular answer, even when they are unsure, and it is this effort that drives students to produce an answer [1]. The utilization of videoconferencing tools during classes brings about a change in student behaviour, resulting in many of them being able to conquer their speech shyness. In virtual settings, students feel more confident and willing to share their opinions [3]. Along with this behaviour of students, during class partitions, teachers maintain a consistent and dynamic communication with their students [4].

Most classes are structured so that the lecturer speaks for at least 80% of the class time [2]. If there are more than 40 students in a classroom, five are expected to be dominant in active participation [2].

Some implemented methods to solve this problem are ARS (Audience Response System) in which a limited number of pre-selected responses (A/B/C/D or true/false) are provided by the interface and audience (students) and they choose one of the answers. But for implementation of this method, the lecture has to be designed in a specific manner to accommodate these questions and cannot include all the topics discussed in the lecture. Another method is chatroom discussion where students chat about or discuss the topic taught in a lecture to help other students, who are confused, understand better. But these chatrooms also encourage discussion which are not related to the lecture and draw students' attention away [2].

A lower level of peer pressure is significantly predicted by an experience of anonymity [5].

In earlier research, it was shown that students may be influenced by their friends and personalities, which may prevent them from expressing their opinions and questions. This may have an impact on their academic performance, and in certain severe circumstances, some students may even give up totally [6]. Researchers have noticed that a combination of classroom learning with online learning can bring about a noticeable change in motivation of students and overall, their interest in study [6]. Additionally, this has an impact on the overall teaching methodology currently in use and causes students to hesitate while studying. Anonymous discussions can assist students in finding clarification and enhancing their learning effectiveness [6]. Researchers have created a system that allows students to text their instructors anonymously and vote on whether they want to ask or answer questions. It was shown that using various methods of contact with teachers could assist students not only overcome their hesitations but also increase their willingness to learn. Therefore, we can analyze that with a good teaching strategy, mobile devices can be a new way of learning tool to use [6]. This way of learning makes learning as a student-centric as most of the students are then responsible for their own learning; he/she must understand how to manage their knowledge and have no boundaries for asking the questions, and it also helps them to improve in their ability to participate in active learning and their ability to comprehend knowledge [6].

5.2 ORDER OF CLASSROOM

The teacher always struggles to strike a balance between two opposing tasks: overseeing the class on the one hand, and keeping an eye on the activities' planned progression and encouraging student participation on the other. A major activity of the classroom is the teacher asking questions and students providing answers and based on the answers the teacher evaluates them. By this 'question-answer-evaluation' the teacher maintains order. The classroom follows a ternary structure of conversation. Questions asked by the teacher are often false questions as he already knows the correct answers. Students learn a whole array of interpretation skills if they participate in classroom evaluation. Teacher and students both interpret each others' behaviour on the basis of activity going in the classroom. The order of the classroom is always in Ethnomethodological framework [7].

5.3 IMPORTANCE OF MAINTAINING ANONYMITY IN CLASSROOM

According to the results, anonymity is a vital factor affecting the willingness of students to participate in class exercises [1]. The advent of technology in education like online classes surely helps teachers but students face challenge in learning by being in the public view of the entire class such as when the teacher asks questions to a student publicly and expecting the right answer or reading the test score of individuals aloud in front of the entire class [8]. Those who are not so bright tend towards introversion, it reduces their confidence. They do not ask question to save themselves from embarrassment by asking so-called "stupid question". They ignore the question which also causes a lack of motivation and poor performance [8].

A report by Paul Love can be found on the Lrnlab Course Website of the Faculty of Education and Social Work at The University of Sydney, in which he says: 'Another benefit of online synchronous communication is that the concept of anonymity allows for greater participation by the students'. Student participation in anonymous discussions can enable them to raise ideas that they would otherwise be hesitant to bring up face-to-face [8]. A student studying at an open university in Israel said he feels more at ease participating in classes online because he doesn't have to reveal his identity [8].

5.4 LEVELS OF ANONYMITY

During the design of platforms that offer anonymity for students in classroom, what levels of anonymity should be provided and what will be its effect on the students and lecturers are often overlooked. According to research by Flinn and Maurer on the importance on anonymity and its categorization [9,10], anonymity can be categorized into six levels.

- Level 5: Authentication and identification occur in a completely secure manner, so there is no possibility that the user can remain anonymous.
- Level 4: Usual identification - Users are recognized within the system by their user name and password, which are required to be entered prior to admittance.

- Level 3: Latent (potential) identification - Users can create and use a variety of distinctive pseudonyms and are referred to as individuals in the system.
- Level 2: The user is identified within the system by a user name and password, but cannot be accurately identified as a person.
- Level 1: Anonymous Identification - Just as they log on anonymously (perhaps using a password), users are identified by the system without a specific identity, without pseudonyms, or with pen names, and the system keeps an event record.
- Level 0: In the case of stand-alone workstations, it is typical to log user activity; however, an application can also do this, as in Level 1.

5.5 SIMILAR WORK

5.5.1 Audience response system

Audience Response Systems for multiple choice and true/false questions have been employed in classrooms in order to address this problem [2]. A limited number of pre-selected responses such as A/B/C/D or true/false are provided by the interface [2]. The lecturer will typically use these interfaces when asking a multiple-choice question directly to the audience. A lecturer's challenge in being effective is anticipating when to reach out to the audience and to structure the lecture specifically so it accommodates this new question-and-answer format. Whether it is student-provided or student-purchased, many of these systems include specialized hardware.

5.5.2 backchan.nl

In a conference or after a talk or after a lecture, members of the audience can organise their collective questions for the speaker using backchan.nl. Most relevant questions are filtered by a moderator from the top-rated questions. Through Conversation Votes, participants are able to indicate consensus by annotating a visual representation of conversation with positive and negative votes. Participants in small groups who were unsatisfied with previous conversations increased their participation after receiving anonymous feedback.

5.5.3 Fragmented social mirror (FSM)

A system with small hand-held devices, with small icons, representing what the feedback was, and a little space to type in text [11]. The text along with icon is then delivered to the lecturer, who then responds to the relevant feedback by the students. During the implementation of this system, usually after the first session, there were also irrelevant feedback by the students, which not only distracted other students, but also the lecturer. The lecturer would laugh along with other students if the feedback was funny and get distracted from the lecture context.

5.5.4 Mobile anonymous question-raising system (MAQ)

With this system, when an instructor is taking a class, a student who has a doubt can ask the question in two ways: first, by selecting among the questions that were asked in past sessions by students who're studying the same course and second, by asking a whole different question that is not present in the list of previously asked questions.

5.6 EXPERIMENTAL RESULTS

5.6.1 Introduction

Providing anonymity to students increases their likelihood to raise their questions. Previous research has shown that common reasons for students to remain silent in the classroom are fear, shyness and anxiety. When students were asked how likely they were to raise their questions in a classroom when they were provided anonymity and when not, most students were more likely to ask questions during class if they were anonymous and in very few cases, the likeliness remained same.

5.6.2 Methodology

The experiment was conducted among 44 college students through Google Forms. The participants had attended an online class at least once before. The form contained two scale-based questionnaires, which students could answer by choosing a number from 1 to 10. The score of 1 represented lowest (or least likely) and 10 represented highest (or most likely). The questions were 'How likely are you to put your doubts forward in an online class?' and 'How likely are you to put your doubts forward in an online class, if you were anonymous to other students?'. The participants were not made aware of the objective of the proposed system and were asked to respond fairly.

5.6.3 Result

In this study, most students were more likely to raise a question during class when they were provided anonymity (Figures 5.1 and 5.2).

As can be observed from the graphs above, if students were not provided anonymity, 23 students (52.27%) were less likely (5 or less on scale) and 21 students (47.72%) were more likely (6 or more on scale) to raise a question during class. If students were provided with anonymity, nine students were less likely and 35 students were more likely to raise questions during class.

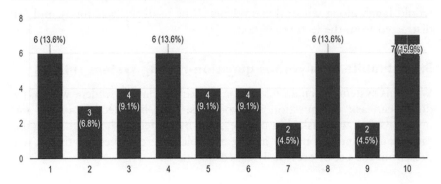

Figure 5.1 Responses of students to when they were not provided anonymity.

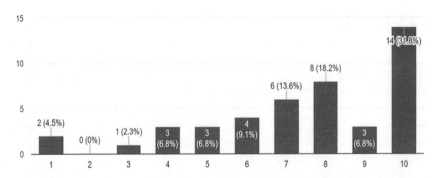

Figure 5.2 Responses of students to when they were provided anonymity.

5.6.4 Discussion

There was a significant increase in the number of students who were more likely to put forward their questions during a class when they were provided anonymity among their classmates. This indicates that making the classroom anonymous among students can bring all the benefits in the teaching learning process from increasing the active participation of students in the classroom.

5.7 CONCLUSION AND FUTURE IMPROVEMENT

It is not easy to make students participate in classroom discussion, even though it is very important for them as well as the lecturer. As opposed to accepting this fact and moving on, technology has stepped in to provide new ways to engage students throughout the lecture. A platform that provides anonymity among students can greatly increase the active participation of students during a class and along with that, make the teaching learning process more effective. But during the design of such a platform, things like extra cost of time and money to set up lectures and teaching should be considered and kept to a minimum. Even though previous similar works have proposed a system and allow students to anonymously give feedback to the lecturer, they are either not cost effective, or do not provide the right level of anonymity, which makes them simply inefficient. The proposed system does not imply any extra cost to the online teaching learning process and is more efficient than previously existing platforms.

The platform provides anonymity to students by hiding the name of participants among the students. This provides anonymity if the students don't recognize each other through voice and it can be further improved by changing the voice of the speaking student with voice-changing technology.

REFERENCES

[1] Freeman, M., Blayney, P., and Ginns, P. (2006). Anonymity and in class learning: The case for electronic response systems. *Australasian Journal of Educational Technology*, 22(4):568–580.

[2] Bergstrom, T., Harris, A., and Karahalios, K. (2011). Encouraging initiative in the classroom with anonymous feedback. In Hutchison, D., Kanade, T., Kittler, J., Kleinberg, J. M., Mattern, F., Mitchell, J. C., Naor, M., Nierstrasz, O., Pandu Rangan, C., Steffen, B., Sudan, M., Terzopoulos, D., Tygar, D., Vardi, M. Y., Weikum, G., Campos, P., Graham, N., Jorge, J., Nunes, N., Palanque, P., and Winckler, M., editors, *Human-Computer Interaction – INTERACT 2011*, volume 6946, pages 627–642. Springer Berlin Heidelberg, Berlin, Heidelberg. Series Title: Lecture Notes in Computer Science.

[3] Sufyan, A., Nuruddin Hidayat, D., Lubis, A., Kultsum, U., Defianty, M., and Suralaga, F. (2020). Implementation of E-learning during a pandemic: Potentials and Challenges. In *2020 8th International Conference on Cyber and IT Service Management (CITSM)*, pages 1–5, Pangkal Pinang, Indonesia, IEEE.

[4] Del Rio-Chillcce, A., Jara-Monge, L., and Andrade-Arenas, L. (2021). Analysis of the use of videoconferencing in the learning process during the pandemic at a University in Lima. *International Journal of Advanced Computer Science and Applications*, 12(5):870–878.

[5] Raes, A., Schellens, T., and Vanderhoven, E. (2011). Increasing anonymity in peer assessment using classroom response technology. In *9th International Conference on Computer-Supported Collaborative Learning (CSCL-2011)*, edited by H. Spada, G. Stahl, N. Miyake and N. Law, vol. 2, 922–923. Hong Kong: International Society of the Learning Sciences (ISLS).

[6] Lai, C.-H., Jong, B.-S., Hsia, Y.-T., and Lin, T.-W. (2020). Use of a mobile anonymous question-raising system to assist flipped-classroom learning. *International Journal of Interactive Mobile Technologies (iJIM)*, 14(03):66.

[7] Paoletti, I. and Fele, G. (2004). Order and disorder in the classroom. *Pragmatics. Quarterly Publication of the International Pragmatics Association (IPrA)*, 14(1):69–85.

[8] Dreher, H. and Maurer, H. (2006). The worth of anonymous feedback. In *Proceedings of the 19th Bled Electronic Commerce Conference (eValues)*. European Commission.

[9] Flinn, B. and Maurer, H. (1996). Levels of anonymity (pp. 35–47). Springer Berlin Heidelberg.

[10] Maurer, H., Calude, C., and Salomaa, A., editors, *J.UCS The Journal of Universal Computer Science*, pages 35–47. Springer Berlin Heidelberg, Berlin, Heidelberg.

[11] Denning, T., Griswold, W. G., Simon, B., and Wilkerson, M. Multimodal Communication in the Classroom: What does it mean for us?. *ACM SIGCSE Bulletin*, 38(1):219–223.

Chapter 6

Speech emotion analyzer using deep learning

Naazneen Ahmed, Ritika Chamaria, Diptarka Paul, Subham Sarangi, Ishika Agarwal, Yamini Sharma, and Srilekha Mukherjee
Department of Computer Science, University of Engineering and Management, Kolkata, India

6.1 INTRODUCTION

Speech Emotion Analysis [1] is an application of machine learning where human speech is recognized by a machine. The machine also recognizes [2] the gender of the speech and the emotion that the person behind the speech is feeling. To understand the working principle of speech recognition and analysis [3] by the machine we must understand what machine learning is along with deep learning. Simply, machine learning is the concept where we feed a bunch of raw data called dataset to the machine system and we train the machine to act a certain way when some inputs are given so that in real life when the environment matches to the machine it will react the same way it is supposed to in the definition given to it by the programmer.

Deep learning [4] is a fragment of the field 'machine learning', such that here artificial neural networks and algorithms are sculpted to perform in a fashion similar to that of the human brain, which learns from a significantly large quantity of data and information. Deep learning is basically numerous layers pertaining to neural networks. These are different algorithms that are loosely modeled to the stated way on how the human brain functions. Training [5] performed with large amounts of data in a huge scale is what actually configures these neurons present in the stated neural network. In the end, the resultant is the required deep learning model that when trained, is able to process new data. Such deep learning models take in data and information from multiple data sources and then analyze that data in real time, without needing any sort of human intervention.

Deep learning is what drives many artificial intelligence technologies that can improve the quality of automation and also analytical tasks. Most people encounter deep learning in their day to day lives when they browse the internet or use their cell phones. Amongst countless other applications, deep learning is utilized to generate captions for YouTube videos, perform speech recognition on mobile phones and smart speakers, provide facial recognition for videos and photographs, and also enables self-driving cars. As researchers and data scientists are handling several increasingly complex projects of deep

DOI: 10.1201/9781003376699-6

learning—leveraging deep learning frameworks—such categories of artificial intelligence will be involved in a greater part of our daily lives. This work also revolves around deep learning where we have to detect the emotion behind human speech using Convolution Neural Network.

6.2 LITERATURE SURVEY

T. Nwe, et al. [6] published a speech emotion recognition based work using the technique of hidden Markov models (HMM). Their works included speech communication and recognition. In the stated paper, some method of text independent speech emotion classification has been proposed. This method makes a proper application of a short time log frequency power coefficients (LFPC) so that it efficiently represents several signals of speech as well as some discrete hidden Markov model (HMM) as a classifier. B. Schuller, et al. [7] published a work on speech emotion recognition that combines several acoustic features as well as various linguistic information within a hybrid support vector machine-belief network based architecture. In the paper, they introduced a very novel method of combining several acoustic features as well as language information with respect to some robust automatic identification/recognition of the speaker's emotion. Seven discrete emotion based states were classified/identified throughout the total work. First, a model for serving the purpose of recognition of several emotion with reference to the acoustic based features/characteristics is presented. These extracted features of several energy, pitch, signal, spectral contours, etc. were given ranks by their individual quantitative contributions with respect to the emotion estimation. There are several other classification methods like that of linear classifiers, neural nets, Gaussian mixture models, and support vector machines, etc., which were compared with this work with respect to the performance. Secondly, another approach to the procedure of emotion recognition done by the stated spoken content is again introduced by giving the belief of network-based spotting for the emotional key-phrases. Thereafter, the two sources of information are finally integrated in a definite soft decision based fusion with the help of a neural net. Finally, the gain is figured out as well as compared to the other several advances. Two different emotional speech corpora that are required for purpose of training along with evaluation are also elaborated in detail. These results that were achieved on applying the stated propagatory novel advances to the recognition of speaker emotion have also been discussed and presented.

K. Lee and J. Lee [8], proposed the work based on the recognition of a noisy speech with the help of a nonstationary ARHMM with the adaptation of a gain under certain unknown noise. In the stated work, ARHMM on mel-scale with some power along with Mel-LPC based time derivative parameters were presented for the purpose of recognition of some noisy speech. These coefficients of mel-scaled AR as well as mel-prediction coefficients for Mel-LPC were calculated on some linear frequency scale from the chosen speech signal. This was done without application of the bilinear transformations on them. Such procedures have been done with the help of a first-order all-pass filter instead of the unit delay. Finally, in addition to this, the Mel-Wiener filter has also been applied to the stated system for the purpose of improving the overall recognition accuracy in the presence of some additive noise. Such a proposed system has been evaluated on the database of Aurora 2. The accuracy of the overall recognition is obtained as 80.02% based on the entire net average.

6.3 PROBLEM STATEMENT

6.3.1 Objective

Two audio datasets have been recorded and fed to the machine learning model. The model is required to detect the classification of emotion being portrayed through audio processing.

The datasets [9] used are as follows:

RAVDESS: This database includes the inclusion of approximately 1,500 audio files from 24 different characters. The 12 female and 12 male characters record short sounds with eight different emotions namely Calm, Happy, Sad, Fear, Surprise, Neutrality, Anger, and Disgust.

Naming audio files is a way for the 7th character to correspond to the different emotions they represent.

SAVEE: This database includes about 500 audio files from four different male characters. The file name has its first two letters as the different emotions it expresses.

This work uses a classification model and appropriate libraries which help to analyze and predict the correct outcome, in this case the emotions portrayed in the datasets.

6.4 PROPOSED SOLUTION

6.4.1 Dataset used

Used Data: The audio data sets were found to contain nearly 2,000 audio files in wav format. The first website contains speech-based data available in three different formats.

1. Audio Visual – Video with speech
2. Speech – Sound only
3. Visual – Video only

We went with the Audio files [10] only because we were dealing with receiving the audio files provided. The zip file contained a total of 1,500 audio files in wav format.

The second website contains about 500 audio speech files from four different players with different types of emotions.

We tested it on one such audio file to determine its features/characteristics by adjusting its waveform (Figure 6.1) and spectrogram (Figure 6.2) as given next.

WAVEFORM:

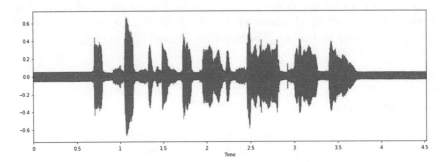

Figure 6.1 Representation of the waveform plot.

SPECTROGRAM:

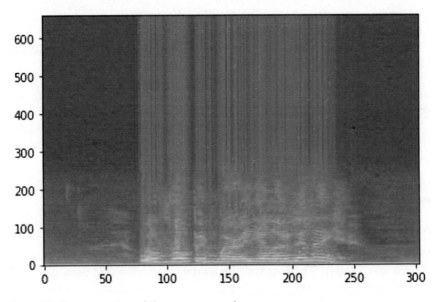

Figure 6.2 Representation of the spectogram plot.

6.4.2 Usage of Librosa package

To analyze and extract features in audio recordings, we used the Librosa package of Python. Librosa is a Python-based tool that analyzes sound and music. It comprises all the components needed to build music acquisition-based systems [11]. We were able to extract features such as MFCC (Mel Frequency Cepstral Coefficient) using the Librosa library. All MFCCs have a common feature of independent speaker and speech identification programs. We also used the IDs provided on the website to distinguish the voices of women and men. This was because, in a study, we found that the distinction between male and female voices can increase by 15%. The effects may also be influenced by the loudness parameter.

Each audio file included several features, which were actually the same type of multiple values. The labels we produced in the previous step were then added to these features. The next stage was to address the lack of features in some audio files that were shorter in duration. To acquire the distinctive qualities of each emotional utterance, we doubled the sample rate. We didn't raise the sample frequency any more since we were concerned that it might accumulate more noise or skew the data.

Getting the features of audio files using librosa

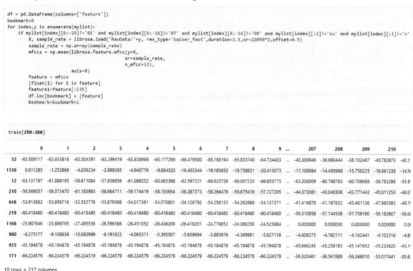

The next steps were to push the data, divide it into train groups and test teams, and design our data training model.

We built the Multi-Perceptron model, the LSTM model, and the CNN model. MLP and LSTM were not suitable due to their low accuracy. CNN has worked well for us because our work is a divisive challenge where we differentiate between different emotions.

6.5 EXPERIMENTAL SETUP AND RESULT ANALYSIS

6.5.1 MLP model

The MLP or Multi Perceptron Model [12] we have created has a very low accuracy of almost 25% authentication in eight layers, softmax output, output size 32 and 550 epochs as seen in Figure 6.3.

MLP

```
In [34]: import numpy as np
         from keras.models import Sequential
         from keras import regularizers
         from keras.layers import Dense, Dropout, Activation, Flatten
         from keras.layers import Convolution2D, MaxPooling2D
         from keras.optimizers import Adam
         from keras.utils import np_utils
         from sklearn import metrics

         num_labels =y_train.shape[1]
         filter_size = 2

         # build model
         model = Sequential()

         model.add(Dense(512, input_shape=(259,),kernel_regularizer=regularizers.l2(0.01),
                     activity_regularizer=regularizers.l1(0.01)))
         model.add(Activation('relu'))
         model.add(Dropout(0.5))

         model.add(Dense(512))
         model.add(Activation('sigmoid'))
         model.add(Dropout(0.5))

         model.add(Dense(num_labels))
         model.add(Activation('softmax'))

         model.compile(loss='categorical_crossentropy', metrics=['accuracy'], optimizer='adam')
```

```
In [105]: plt.plot(history.history['acc'])
          plt.plot(history.history['val_acc'])
          plt.title('model accuracy')
          plt.ylabel('accuracy')
          plt.xlabel('epoch')
          plt.legend(['train', 'test'], loc='upper left')
          plt.show()
```

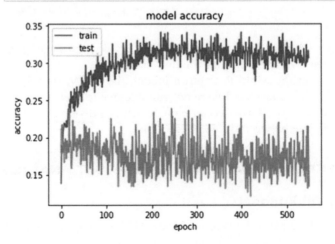

Figure 6.3 Representation of the accuracy plot with respect to MLP.

6.5.2 LSTM model

LSTM [13,14] has tan h performance function and low training accuracy of about 15% in five layers, 32 size collections and 50 epochs (Figure 6.4).

LSTM

```
In [35]: from keras.layers import LSTM
         from keras import optimizers
         modellstm = Sequential()
         modellstm.add(Embedding(1547, 259))
         modellstm.add(LSTM(400,dropout=0.10,return_sequences=True))
         modellstm.add(Dense(256, activation='softmax',kernel_regularizer=regularizers.l2(0.001),
                      activity_regularizer=regularizers.l1(0.001)))
         modellstm.add(LSTM(128,dropout=0.3))
         modellstm.add(Dense(8, activation='tanh'))

         modellstm.compile(loss='categorical_crossentropy',optimizer='SGD',metrics=['accuracy'])
         #opt = optimizers.SGD(lr=0.0001)
         #modellstm.compile(loss = "categorical_crossentropy", optimizer = opt,metrics=['accuracy'])

In [38]: modellstm.summary()
```

Layer (type)	Output Shape	Param #
embedding_1 (Embedding)	(None, None, 259)	400673
lstm_1 (LSTM)	(None, None, 400)	1056000
dense_4 (Dense)	(None, None, 256)	102656
lstm_2 (LSTM)	(None, 128)	197120
dense_5 (Dense)	(None, 8)	1032

Total params: 1,757,481
Trainable params: 1,757,481
Non-trainable params: 0

6.5.3 CNN model

The best model [15] for our division was the CNN model. We found the best 60% authentication with 18 layers, rmsprop refresh function, softmax activation function, 32 size size and 1,000 times, after training multiple models (Figure 6.5).

```
In [91]: plt.plot(lstmhistory.history['acc'])
         plt.plot(lstmhistory.history['val_acc'])
         plt.title('model accuracy')
         plt.ylabel('accuracy')
         plt.xlabel('epoch')
         plt.legend(['train', 'test'], loc='upper left')
         plt.show()
```

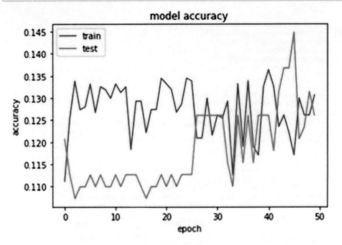

Figure 6.4 Representation of the accuracy plot with respect to LSTM.

```
In [183… plt.plot(cnnhistory.history['loss'])
         plt.plot(cnnhistory.history['val_loss'])
         plt.title('model loss')
         plt.ylabel('loss')
         plt.xlabel('epoch')
         plt.legend(['train', 'test'], loc='upper left')
         plt.show()
```

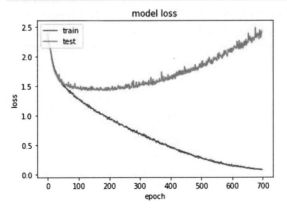

Figure 6.5 Representation of the accuracy plot with respect to CNN.

```
model.summary()
Model: "sequential_4"
_____
Layer (type)                 Output Shape              Param #
=================================================================
conv1d_20 (Conv1D)           (None, 216, 256)          1536
_____
activation_24 (Activation)   (None, 216, 256)          0
_____
conv1d_21 (Conv1D)           (None, 216, 128)          163968
_____
activation_25 (Activation)   (None, 216, 128)          0
_____
dropout_6 (Dropout)          (None, 216, 128)          0
_____
max_pooling1d_4 (MaxPooling1 (None, 27, 128)           0
_____
conv1d_22 (Conv1D)           (None, 27, 128)           82048
_____
activation_26 (Activation)   (None, 27, 128)           0
_____
conv1d_23 (Conv1D)           (None, 27, 128)           82048
_____
activation_27 (Activation)   (None, 27, 128)           0
_____
conv1d_24 (Conv1D)           (None, 27, 128)           82048
_____
activation_28 (Activation)   (None, 27, 128)           0
_____
dropout_7 (Dropout)          (None, 27, 128)           0
_____
conv1d_25 (Conv1D)           (None, 27, 128)           82048
_____
activation_29 (Activation)   (None, 27, 128)           0
_____
flatten_4 (Flatten)          (None, 3456)              0
_____
dense_4 (Dense)              (None, 10)                34570
_____
activation_30 (Activation)   (None, 10)                0
=================================================================
Total params: 528,266
Trainable params: 528,266
Non-trainable params: 0
```

After training the model we had to predict the emotions in our test data. The following picture shows our predictions and real numbers.

```python
model.add(Conv1D(256, 5,padding='same',
                 input_shape=(216,1)))
model.add(Activation('relu'))
model.add(Conv1D(128, 5,padding='same'))
model.add(Activation('relu'))
model.add(Dropout(0.1))
model.add(MaxPooling1D(pool_size=(8)))
model.add(Conv1D(128, 5,padding='same',))
model.add(Activation('relu'))
model.add(Conv1D(128, 5,padding='same',))
model.add(Activation('relu'))
model.add(Conv1D(128, 5,padding='same',))
model.add(Activation('relu'))
model.add(Dropout(0.2))
model.add(Conv1D(128, 5,padding='same',))
model.add(Activation('relu'))
model.add(Flatten())
model.add(Dense(10))
model.add(Activation('softmax'))
opt = keras.optimizers.RMSprop(lr=0.00001, decay=1e-6)
```

In [218...] `finaldf[:10]`

Out[218...]

	actualvalues	predictedvalues
0	female_angry	female_fearful
1	female_happy	female_fearful
2	male_fearful	male_sad
3	female_sad	female_happy
4	male_angry	male_angry
5	female_angry	female_angry
6	male_fearful	female_angry
7	male_happy	male_angry
8	male_angry	male_fearful
9	female_fearful	female_angry

6.6 CONCLUSION AND FUTURE SCOPE

After working with various models, the best model that suited our emotion categorization problem is the CNN model which provides us with the highest accuracy. We achieved the highest valid accuracy of 50.76% with our model. The performance of our model will improve more if the supply of data is vast in size to work with. The essential part about our model is that it can distinguish the male and female voices very well. Our model predictions as opposed to actual values can be analyzed in the above chart.

In the beginning, the accuracy of our model was around 30%, then it rose to 50% and hopefully, it will improve more in the future. With that, here are some of the fields which can be improved with the application of our model. Firstly, the recommendations of the product in online shopping portals will become more accurate in matching the mood of the customers. Secondly, lie detectors will be able to analyze the minute variations in the speech and also the emotional state of the subjected person more profoundly. And lastly, it will help in communicating with machines. Though machines can understand each and every word, it is hard for them to understand the consciousness behind those words. This deficiency can be rectified to a large extent by the application of our model.

REFERENCES

[1] Kerkeni, L., Mbarki, M., Raoofand, K., et al. (2018), Speech emotion recognition: Methods and case study, ICAART.
[2] Singh, A., Srivastava, K.K., and Murugan, H. (2020), Speech emotion recognition using convolution neural network (CNN). Vol. 24.
[3] Sundarprasad, N. (2018), Speech emotion detection using machine learning techniques, Master's Projects. 628.
[4] Abbaschian, B.J. (2021), Deep learning techniques for speech emotion recognizer, Special Issue Sensors for Rehabilitation, Telemedicine and Assistive Technology.
[5] Miteshputhranneu (2021), Analysis and work on speech emotion analyzer, PhD Thesis.
[6] Nwe, T., Foo, S., and De Silva, L. (2003), Speech emotion recognition using hidden Markov models, *Speech Communication*, vol. 41, pp. 603–623.
[7] Schuller, B., Rigoll, G., and Lang, M. (2004), Speech emotion recognition combining acoustic features and linguistic information in a hybrid support vector machine-belief network architecture, *Proc. ICASSP 2004*, vol. 1, pp. 577–580.
[8] Prudhvi GNV (2020), Guide for speech emotion recognition using deep learning, *An Analyzer Guide*.
[9] Wu, S., and Falk chan, W.Y. (2010), Automatic speech emotion recognition, speech communication.
[10] Cano, A. (2020), Social media and machine learning. pp. 1–96.

[11] Khalil, R.A. (2019), Speech emotion recognition using deep learning techniques- A Review, *IEEE Access*, vol. 7.
[12] huang, C. (2014), A research of speech emotion recognition based on deep belief network and SVM, *Mathematical Problems in Engineering, Hindawi*.
[13] Nicholson, J., Takahashi, K., and Nakatsu, R. (2000), Emotion recognition in speech using neural networks, *Neural Computing & Applications*, vol. 9, pp. 290–296.
[14] Poritz, A. (1982), Linear predictive hidden Markov models and the speech signals, *ICASSP*, pp. 1291–1294.

Chapter 7

Technology-enhanced personalized learning in higher education

Ravi Kant Verma[1], Satyendra Gupta[2], and Svitlana Illinich[3]

[1]Research Scholar, School of Education, Galgotias University, G.B. Nagar, U.P. India
[2]Professor and Dean, School of Education, Galgotias University, G.B. Nagar, U.P. India
[3]Associate Professor, Department of Social Technologies, Vinnytsia Institute and College of Open International University of Human Development, Vinnytsia Oblast, Ukraine

7.1 BACKGROUND OF PERSONALIZED LEARNING

Personalized learning is not a novel idea. For many years, many schools and teachers have successfully modified their curricula and teaching strategies to meet the needs of kids and teenagers. The initiative to make best practices universal is what is new. The foundation for every school becoming great is reimagining the educational system around young people's learning requirements and talents. To create a successful personalized learning system, we must first acknowledge that every child deserves the opportunity to achieve their full potential, regardless of talent or background. Personalized learning entails excellent instruction that is sensitive to the various ways in which students succeed. Pursuing this course of action makes sense from a moral and educational perspective. In addition to producing excellence, a system that responds to each student by designing an educational path that takes into account their needs, interests, and aspirations will also significantly advance equity and social justice.

Personalization, almost equivalent to student-centred learning, is heavily focused on the individual student's interests, requirements, and goals. By integrating them into the design, choosing, and execution of the course content, student-centred learning aims to keep students interested in their studies [1].

For many years, the terms customized learning and customization in education have been used interchangeably. There are several ways to define it. According to several sources, including Basye [2], the term "personalized learning" was used to market more technology tools and programmes intended to provide students with personalized teaching, but it really fell short of meeting students' educational needs.

Technology has enabled the development of personalized learning for pupils, which is used to assist students in setting and achieving objectives, engaging in their education, taking into account their interests and aspirations, and creating plans to accomplish those goals. Personalization in the classroom requires the teacher to prioritize the needs of the students,

DOI: 10.1201/9781003376699-7

adapt their instruction to fit their individual learning styles, and work with the assistance of the staff, administration, and peers. Project-based learning, community-based learning, blended learning, and authentic learning are some of the various educational approaches and instructional methods used to achieve this goal. It serves as an alternative to conventional approaches to education, which let teachers act alone and with little to no involvement from pupils [3,4].

Patrick et al. [5] claim that by encouraging students to set specific goals and expectations for success, personalizing learning helps students make informed decisions in a demanding and challenging learning environment. In this environment, teachers have the time they need to connect with students, provide challenging, adaptable, and flexible training, and place a strong focus on critical thinking and metacognitive abilities to promote deeper, more in-depth, and more independent learning.

In order to retain student engagement, achieve academic success, and become ready for future educational possibilities, personalized learning takes into account each student's interests, voice, choice, and requirements. Students take part in the planning process directly and have access to their learning plans at home or school to better understand their progress towards attaining their objectives and decide what additional needs to be done or developed in light of their future goals and paths to success [6,7].

Personalized learning theory is a complicated educational strategy that incorporates objectives to meet individual learning requirements. It is described by the U.S. Department of Education as education that is adapted to meet the individual learning requirements of each student, differentiated according to their preferred learning style, and taken into account their interests and learning objectives [8].

Personalized learning is an opportunity for educators to build learning opportunities that take advantage of the digital capabilities that the majority of students currently have. Customized learning is individualized for every learner's abilities, needs, and interests while maintaining the highest attainable standards. This technique represents a significant departure from the usual "one-size-fits-all" approach to education. Personalization urges instructors to be more open and flexible in order for students to be more committed to building their own individualized learning paths. Students who participate in individualized learning at their own pace are provided tools and feedback that encourage them to maximize their particular abilities and potential.

The term "personalized learning" refers to the plethora of educational programmes, teaching strategies, and academic assistance tools created to satisfy each student's unique learning requirements. By evaluating each student's learning needs, interests, and aspirations before providing personalized instruction, personalized learning aims to help every student succeed academically. Each student should have a voice in decision-making over their education, including what and how they wish to learn. There are many different ways to conceptualize the personalization of learning.

- Content personalization: students interact with material, topics, and geographic regions that particularly interest them.
- Tempo and progress customization learning, in which students move through the subject matter and grade levels at their own pace.
- Process personalizing learning: depending on the needs and interests of the pupils, different instructional strategies and learning settings are used.

7.2. COMPONENTS OF PERSONALIZED LEARNING

According to [9,10], personalization must include the following essential elements. Figure 7.1 shows the components of personalized learning.

1. Student organizations
2. Personalized instruction
3. Each student receives immediate instructional interventions and support as needed
4. Adjustable pacing
5. Specific pupil profiles
6. More thorough education and meaning-making through problem-solving

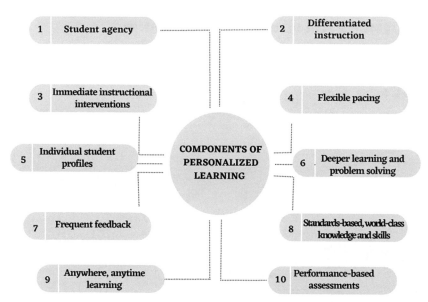

Figure 7.1 Showing components of personalized learning [9].

7. Regular feedback from teachers and fellow students
8. World-class, standards-based knowledge and abilities
9. Learning anytime and anywhere
10. Evaluations based on performance.

7.3 FUNDAMENTAL PRINCIPLES OF PERSONALIZED LEARNING

According to [11] several educational institutions are attempting to develop upon the core ideas of personalized learning, which put the student first, rather than the instructor or the curriculum. All students should participate in the following activities as part of their personalized learning. See Figure 7.2.

- Learner profiles that detail the particular abilities, gaps, shortcomings, strengths, interests, and ambitions of every pupil.
- Customized learning routes with goals and objectives for each student, as well as a variety of learning opportunities catered to their requirements;
- Individual mastery, which entails continuously evaluating progress.
- Flexible learning environments that may include various methods of delivering instruction and continuously maximizing the use of resources to support student learning.

7.4 DESIGNS FOR PERSONALIZED LEARNING

Attendees of the Invent to Educate: System [Re] Design for Customized Learning conference in 2010 identified five crucial components of customized learning as per Figure 7.3:

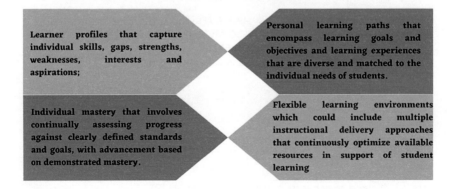

Figure 7.2 Showing fundamental principles of personalized learning [11].

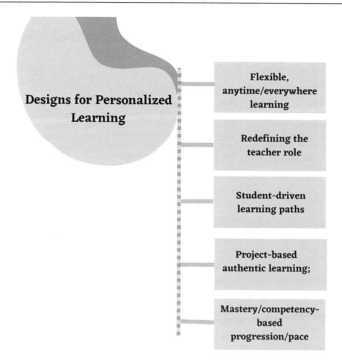

Figure 7.3 Showing designs for personalized learning [12].

- Adaptable, anywhere/any moment education;
- Redefining the position of the teacher;
- Real-world project-based education;
- Learning pathways driven by students; and
- Pace/progression dependent on mastery/competence [12,13].

7.5 MODERN PEDAGOGY CENTRED ON PERSONALIZED LEARNING

By avoiding any learning obstacles and concentrating teaching and learning on students' aptitudes and interests, these ideas aim to enhance standards. How to collaboratively construct this offer for each kid and each parent is the major problem, the fundamental components of individualized education both within and outside of the classroom according to Rooney, Brown, Sommer, & Lopez [14]. The following modern pedagogy, as shown in Figure 7.4, is centred on personalized learning.

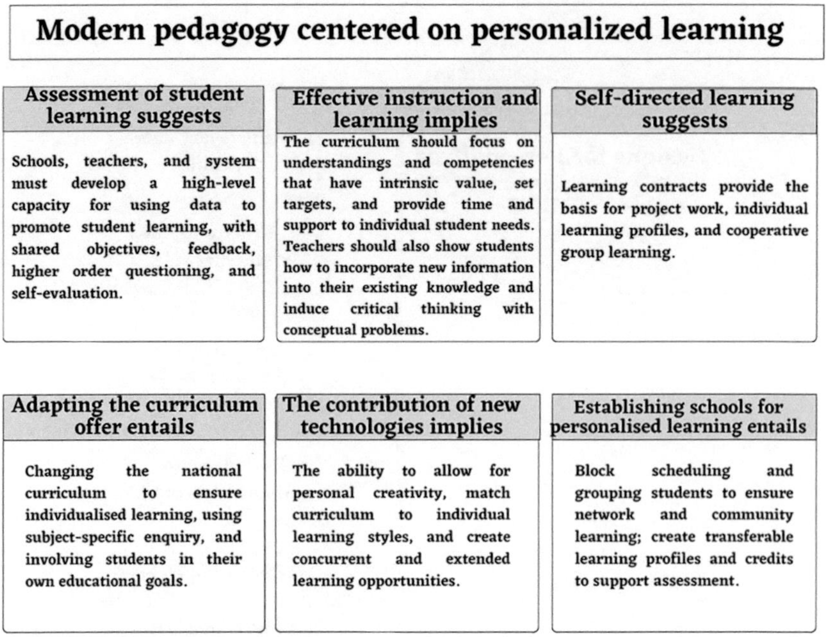

Modern pedagogy centered on personalized learning

Assessment of student learning suggests	Effective instruction and learning implies	Self-directed learning suggests
Schools, teachers, and system must develop a high-level capacity for using data to promote student learning, with shared objectives, feedback, higher order questioning, and self-evaluation.	The curriculum should focus on understandings and competencies that have intrinsic value, set targets, and provide time and support to individual student needs. Teachers should also show students how to incorporate new information into their existing knowledge and induce critical thinking with conceptual problems.	Learning contracts provide the basis for project work, individual learning profiles, and cooperative group learning.

Adapting the curriculum offer entails	The contribution of new technologies implies	Establishing schools for personalised learning entails
Changing the national curriculum to ensure individualised learning, using subject-specific enquiry, and involving students in their own educational goals.	The ability to allow for personal creativity, match curriculum to individual learning styles, and create concurrent and extended learning opportunities.	Block scheduling and grouping students to ensure network and community learning; create transferable learning profiles and credits to support assessment.

Figure 7.4 Showing Modern pedagogy centred on personalized learning [14].

7.5.1 Assessment of student learning suggests
- The development of a highly developed capacity for the system, instructors, and school to use data to promote student learning.
- The process of gathering and examining data to assist educators and students in making choices about where kids are in their learning, where they need to go, and the most effective route to get there; and
- Mutually agreed-upon objectives, constructive criticism that points out potential improvement areas, higher-order inquiries, and peer and self-evaluation.

7.5.2 Effective instruction and learning imply
- Long-lasting and intrinsically valuable information and skills should be prioritized in the curriculum.
- While everyone should have high expectations and demanding goals, time and assistance should vary depending on the requirements of each individual student but standards should always be the same; and
- To show students how to integrate new knowledge into their past understanding, teachers should utilize conceptual questions to elicit critical thinking in their classes.

7.5.3 Self-directed learning suggests

- The project work, an important and continuous aspect of the curricular offer, is built around these learning contracts.
- Personal learning profiles,
- Contrary to expectations, cooperative group learning and social interaction are strongly emphasized.

7.5.4 Adapting the curriculum offer entails

- Altering the structure of the national curriculum to provide continuity for customized learning across the three levels of education – foundational, middle, and 14–19 years old;
- Establishing standards-based subject-specific enquiry as the cornerstone of curricular delivery; and
- Allowing pupils to design their own educational objectives. This is necessary for children to cultivate a love of learning and a commitment to their education over the long run.

7.5.5 The contribution of new technologies implies

- The capacity to accommodate individual inventiveness, the capacity to adapt the curriculum to different learning preferences, and the capacity to provide the learner control over the rate of learning;
- Opportunities for concurrent and extended learning outside of the typical school day; and
- Creating diagnostic tests for learning with several directions to go.

7.5.6 Establishing schools for personalized learning entails

- Providing each learner with a connection to an adult; differentiating the workforce for student learning; and enhancing the function of the learning mentor.
- Block scheduling and grouping students in accordance with their learning requirements, both inside and between schools, to ensure network and community learning.
- Establishing a system of credits and transferable learning profiles to assist assessments and provide flexibility.

Each of these six components, as shown in Figure 7.4, is obviously essential to developing a contemporary pedagogy that is centred on learning, but they are not necessarily incompatible with one another. To ensure that every student reaches their potential concentrate on two areas of personalization. The first step entails adjusting the curriculum to allow for a variety of subject matter, personal relevance, and flexible learning pathways to be provided across the educational system. The second is met cognition, sometimes known as learning how to learn.

7.6 MOTIVATION FOR PERSONALIZED LEARNING

According to Tranquillo and Stecker [15], the learning process is greatly impacted by motivation. While some individuals get more knowledge from external sources, others could accomplish more by following their own goals. Everyone participating in any learning process, regardless of circumstance, should be aware of how motivation influences learning. There are two types of motivations.

First, extrinsic motivation and second, intrinsic motivation

Extrinsic motivation arises when we are driven to engage in an action or activity in order to receive a reward or avoid punishment.

Intrinsic motivation entails participating in conduct that someone personally finds fulfilling; simply, doing something for the purpose of doing it rather than the desire for some external benefit.

Table 7.1 describes the differences between extrinsic motivation and intrinsic motivation.

7.6.1 Extrinsic motivation for personalized learning

According to Cherry [16,17], the extrinsic motivation of personalized learning is discussed below and shown in Figure 7.5.

- **Instrumental** – In terms of education, the industrial model, which promotes rewards and consequences, is analogous to the instrumental system. We find it tough to change since we are accustomed to this system. While some students are driven to adhere to the rules, others lack motivation owing to a lack of interest, failure, or boredom with education.

Table 7.1 Difference between extrinsic motivation and intrinsic motivation

Extrinsic motivation	Intrinsic motivation
1. Engaging in sports in order to gain awards.	1. Attempting to play a sport because you like the activity.
2. To avoid being chastised by your parents, clean your room.	2. Clean your room because you like cleaning.
3. Participating in a scholarship competition.	3. Completing a word puzzle because the challenge appeals to you.
4. Studying in order to earn a high grade.	4. Learning a subject that interests you.

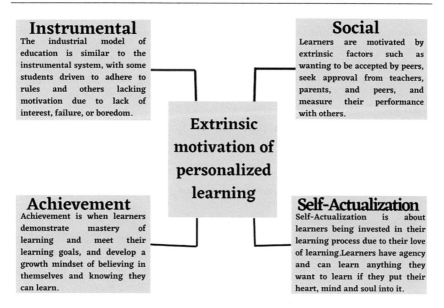

Figure 7.5 Showing extrinsic motivation of personalized learning by Cherry [16].

- **Social** – Learners are motivated by extrinsic factors such as wanting to be accepted by peers, seeking approval from teachers, parents, and peers, and measuring their performance with others. Friends mean more to them than school, but they are still motivated to learn due to extrinsic motivations.
- **Achievement** – Accomplishment is when students show that they are motivated to study and that they want to do well in school. They adopt a growth mindset, which is the belief that they can learn, and they select evidence that shows learning mastery and how they achieved their learning objectives.
- **Self-Actualization** is the process through which students become involved in their education because they like studying. The focus is entirely on the individual and how they choose to learn, whether it be by picking up a new ability, expanding their knowledge, building something they never imagined they could, or following their goals. Learners who are self-actualized have agency and understand that everything they set their minds to can be learned if they put their all into it.

7.6.2 Intrinsic motivation for personalized learning

According to Rickabaugh [18], a method of teaching and learning is based on the requirements, interests, and preparedness of each individual learner. The learners have an active role in establishing objectives, organizing learning routes, monitoring development, and choosing how learning will be shown. When learners pursue competency in line with defined standards, learning objectives, material, techniques, and pace are likely to differ from learner to learner at any given time. A learning environment that is completely individualized goes beyond individualization and differentiation.

Consider that the learner performs the following, as per personalized learning:

- Is self-motivated
- Makes connections between learning and interests, abilities, passions, and ambitions
- Participates actively in the planning of their education
- Is responsible and accountable for their education, including their voice and choice in what they study and how.
- As the learner moves along their learning path with instructor guidance, they set goals and benchmarks for their learning plan.
- Develops the capacity to select and apply resources and technologies that will assist and promote their study.
- Establishes a network of peers, experts, and educators to encourage and support their academic pursuits.
- Exhibits domain competence within a competency-based framework.
- Develops into a self-motivated, experienced learner who monitors their progress and assesses their education in light of their level of knowledge and competence.

This section looks at the seven elements of learner agency as provided by Bray and McClaskey [19–22] in their book How to Customize Learning: A Practical Guide. Figure 7.6 shows the intrinsic motivation of personalized learning.

7.7 THE STAGES OF PERSONALIZED LEARNING ENVIRONMENTS

Miliband [23–25] defines personalized learning as having five stages shown in Figure 7.7.

- One of the early phases is the evaluation phase when teachers and students work constructively together to identify areas of strength and weakness.
- The ability to choose learning methodologies is given to instructors and students in the second stage, which entails teaching and learning.
- The third level, known as a curricular choice, allows students to choose their own course of study, fostering student autonomy.

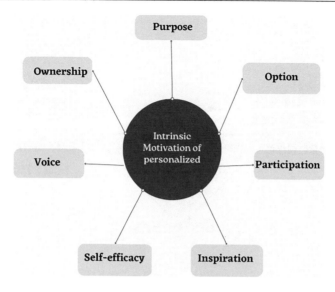

Figure 7.6 Showing intrinsic motivation of personalized learning by Bray and McClaskey [19].

Stages of personalized learning

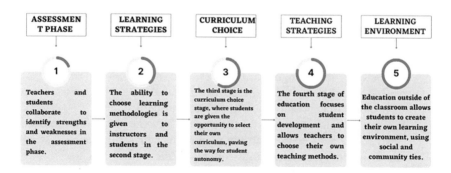

Figure 7.7 Showing stages of personalized learning environments by Miliband [23].

- The fourth stage represents a major break from the conventional educational paradigm since it is focused on student development and gives teachers the freedom to select their own teaching methods.

The last level of education involves studying outside the traditional classroom, which gives students the freedom to choose their own learning environment. During this phase, social and community ties are utilized.

According to Barbara Bray and Kathleen McClaskey (2013), there are three stages of personalized learning as shown in Figure 7.8.

82 Advances in technological innovations in higher education

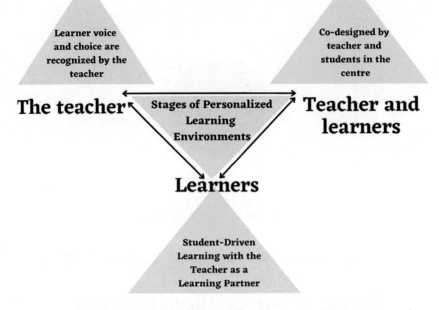

Figure 7.8 Showing stages of personalized learning environments (McClaskey, 2013).

1. **Learner voice and choice are recognized by the teacher**
   The teacher:
   - Recognizes how each student best learns and bases instructional decisions on strengths, difficulties, and interests.
   - Redesigns the classroom environment and creates courses and projects that promote learner voice and choice.
   - Thoughtfully incorporates technology into the curriculum in order to educate all learners in the manner in which they learn best.
2. **Co-designed by teachers and students in the centre**
   The teacher and learners:
   - Select the knowledge-based tools they will use to support their learning;
   - Co-design lessons and projects with learners to include their voice and choice;
   - Know how to select and use the acceptable tools to support learning;
   - Are moving towards a performance scheme where students begin to show that they have mastered the material.
3. **Student-Driven Learning with the Teacher as a Learning Partner**
   Learners:
   - Direct their education in accordance with their interests, goals, and inquiries.

- Learn at their own speed with a continuous feedback loop, monitoring their development.
- Create adaptable projects that let students express themselves and decide how to demonstrate proficiency in competency-based learning.

7.8 THE ROLE OF TECHNOLOGY IN PERSONALIZED LEARNING

Helping students study at their own speed is one of the top technologies-based customized learning benefits. Also, thanks to the digital format, teachers may adapt the curriculum to the ability of their pupils. Radical changes in how education is provided and received are being brought about by the rise of technology-based personalized learning. With the digitization of education, educational programmers' reach is expanding daily. Technology-based personalized learning is giving instructors and students new ways to teach and learn, encouraging increased student involvement in the learning process as a whole.

The way education is delivered in schools and universities has changed with the introduction of new technology-aided learning tools like smart boards, MOOCs, tablets, laptops, etc. One of the most economical methods to teach young brains is through the Internet of Things, which is further proving to be so. Also, it is an effective approach for including a top-notch educational opportunity for everyone. The edutech industry is always looking for new ways to improve access to education for those who are now unable to afford quality educational facilities [26,27].

Young people from all around the world are increasingly reporting having attended online courses in the past. The wonderful thing about digital education is that you can produce things once and reuse them over and over again for future generations. Digital education also gives teachers the freedom to alter their lesson plans to meet the specific needs of each student.

The following Figure 7.9 shows the role of technology in personalized learning as per the "Office of Educational Technology" [26].

7.9 PERSONALIZED LEARNING ENVIRONMENT TOOLS AND SOLUTIONS

In India, digital education is transforming the way pupils study several disciplines. Current Internet technologies and artificial intelligence enable students' learning experiences to be tailored to their individual time constraints and demands. Furthermore, a wide range of digital learning solutions and technologies are now available.

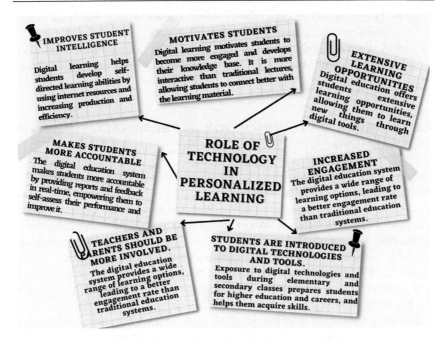

Figure 7.9 Showing the role of technology in personalized learning.

Several instructors have used Online services to create websites for their students, especially when there was no web hosting service available or when there were limitations on web posting that made it difficult for teachers to give their pupils access to the necessary information. The curricula should be customized by educators, institutions, and language businesses for certain student groups. New homepage services provide more capability and flexibility than older ones. One of the fundamental ideas of a Personalized Learning Environment (PLEs) is the capacity to include flexibility. Wikis, forums, Twitter, social networks, bookmaking, and other readily available tools and services are all intended to encourage the distribution of the material produced by individual users through participation, engagement, and collaboration. These technologies, as shown in Table 7.2, could aid in enhancing performance and engagement through personalized learning.

7.10 SCHOOL AND DISTRICT LEADERS CAN SUPPORT PERSONALIZED LEARNING

School and district administrators need to make sure that the appropriate digital resources, technological infrastructure, and classroom and school-wide procedures are in place to lessen the potential burden of individualized learning on instructors. This involves researching and selecting curriculum, materials,

Table 7.2 Showing personalized learning environment tools

Tool	Services	Tool	Services
Address books	Address books	Microblogs	Twitter
Analytics	Google Analytics	Music	Music
Blogs	Blogger, WordPress	Online office	Google Docs
Books store	Amazon	Other media	TV, radio
Calendar	Calendar	Physical	objects/sites, libraries, books, etc.
Chat/IRC	MSN	Podcast	Podcast
Courses	Free courses	Portfolios	e-portfolios
Curriculum documents	Curriculum documents	RSS	RSS Reader, Feedly, Blog lines
Databases	Databases	Search engine	Google, Yahoo
Drawings	Drawings	Slide casting	SlideShare
Email	email	Social bookmarking	Delicious, Diigo, Digg, Zotero
Files/document repository	Dropbox, Drive	Social networks	Facebook, LinkedIn, Academia, Plurk, Elgg, Base camp
Fora	Fora	Start pages	Netvibes, Protopage, iGoogle, Windows Live (discontinued)
Image sharing	Flicker	Video/repository	YouTube
LMS	Module, Sakai, Blackboard	Videoconference	Skype, Flash Meeting
Mail lists/news	Mail lists/newsletters	Virtual worlds	Second Life
Maps	Google Maps	Webinars	Webinars
Wikis	Wiki spaces	Wikipedia	Wikipedia

tools, and other technological resources for education, as well as assisting instructors in providing substantial feedback and formative assessment [26].

Data-informed personalized learning may be supported by school and district leaders by:

- To enhance instruction and learning, teachers should leverage learning environments to provide students access to data.
- Teachers should have time to think about data and plan for its application.
- Investing in teachers' education, mentoring, and ongoing professional development.
- Fostering opportunities for informal cooperation among educators and institutions to talk about best practices.

- Establishing conditions in which educators are backed up as they design and personalized instruction for different students.
- Supplying and keeping up with the technology necessary for personalized instruction.
- Investing in infrastructure, such as cutting-edge tools and contemporary data systems.
- Holding dialogues with local residents, parents, and teachers about the benefits of data-informed education.

7.11 STATE SUPPORT FOR PERSONALIZED LEARNING

Governments can also provide more direct assistance to schools and teachers as they transition to and sustain personalized learning activities. Office of Educational Technology [26] makes the following recommendations to state leaders:

- Explain and give advice on state rules that may have an impact on personalized learning plans.
- Provide financing channels to support the technologies required for personalized learning.
- Examine how teacher education programmes approach personalized learning.
- Make it accessible to school districts, teachers, and students as important and pertinent tailored learning materials.
- Invest the required human capital resources in making sure that state authorities are informed about and supportive of customized learning initiatives.
- Keep track of the state's results for individualized learning.

7.12 EFFECTIVE PERSONALIZED LEARNING

Learning is an ongoing process in which no two pupils have the same manner or speed of learning. Personalized learning provides dynamic settings that foster a love of learning by adapting the curriculum tailored to each student's ability of success and skills.

- You've probably heard of the proximal development zone, which claims that everyone has a set of abilities that can be improved with assistance. These abilities are considered proximal because the individual requires instruction and nurture to acquire the capacity to accomplish these tasks on their own.
- Personalized learning is built on similar ideas to the zone of proximal development. It strives to set difficult but feasible goals for kids based

on their unique needs and talents. Personalized learning also entails instructors actively reviewing student progress, interacting with parents, and giving further assistance when necessary.
- Personalized learning leads to higher productivity and material delivery as students require it. This implies that content that is already familiar may be omitted, and more time can be spent on topics and places that are proving difficult. Teachers spend less time re-teaching subject that is already familiar as a consequence of individualized learning. Students are better able to focus on areas that demand attention and offer the knowledge required in a continuous learning journey.
- The key to effective personalized learning is a data-driven strategy that takes into account the demands of each individual learner. The correct digital solutions may help educators track and monitor progress more easily, giving them the knowledge, they need to make educated decisions about how to best support each individual student.

Students may develop at their own pace as a consequence of this strategy, which can help them stay interested in their studies. Using data from management of learning and other sources such as Google Analytics or Module, teachers can diversify instruction and tailor it to the requirements of each student.

7.13 PERSONALIZED LEARNING IS CHANGING EDUCATION

Personalized learning is a popular topic in education right now. That is not to say personalized learning is a bad thing; rather, it is a hot issue with a lot of consequences; there is a lot to speak about when it comes to personalized learning. As a result, we must be specific. What kind of personalized learning are we discussing?

- Pupils can achieve mastery learning, which is entire and full knowledge of a certain subject, topic, or concept.
- Students benefit from one-on-one mentoring, which consists of frequent meetings with the same teacher.
- Online tutoring is inexpensive. Online tutoring is more economical for most families than other traditional and contemporary individualized learning choices. In addition, online tutoring is substantially less expensive than in-person tutoring.
- Online instruction is adaptable.
- Online tutoring is available; students may plan sessions when and when they wish. Students may now access top-tier instructors via their mobile devices, such as IOS or Android smartphones or tablets, or from PCs, unlike ever before. Geography is no longer a barrier to tailored learning in this concept.

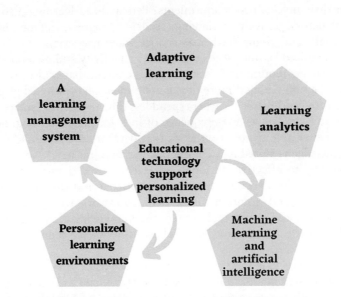

Figure 7.10 Showing educational technology supports personalized learning.

7.14 EDUCATIONAL TECHNOLOGY SUPPORT PERSONALIZED LEARNING

Educational technology may help with tailored learning in a variety of ways shown in Figure 7.10.

Adaptive learning technology modifies the content and complexity of learning material based on a student's progress and performance. This allows students to study at their own speed and concentrate on the issues that require the greatest assistance.

Management of learning is a platform that allows students and teachers to access course materials, submit assignments, and interact with one another. It may also monitor a student's development and give comments and assistance.

Learning analytics software may track a student's learning progress and give instructors and students data-driven insights. This can assist teachers to adjust their lessons to particular student's requirements.

Personalized learning environments: These are digital platforms that allow students to set their own learning paths and goals. They may also allow students to personalize their learning experiences by allowing them to choose their own learning resources or establish their own learning pace.

Artificial intelligence and machine learning: AI and machine learning may be used to deliver tailored learning suggestions based on a student's interests and requirements. An AI-powered tutoring system, for example, may be able to deliver personalized lessons and feedback depending on a student's skills and deficiencies.

Eventually, educational technology has the potential to be a significant instrument for encouraging personalized learning by providing students with customizable and flexible learning experiences that are matched to their unique requirements and goals.

7.15 IMPORTANCE OF PERSONALIZED LEARNING

Several characteristics define the human personality. A unique format is created by a combination of particular characteristics. We must acknowledge that everyone is built differently and that there must be a varied approach to teaching. If we look closely, we will notice that each student's learning needs differ from those of the others. A learner may spend hours learning a topic, whilst another may just need a few minutes. Some people are better at explaining things than others at writing.

More student engagement results from personalized learning. We will increase and promote student engagement by addressing the various interests of students. Our kids will spend more time learning and truly grasp the subject if they are more engaged. What more could a teacher ask for? Increased involvement also results in higher levels of motivation among pupils. Students will get involved in their learning journey because they have a level of choice in the learning path they pursue. This incentive will help them avoid distractions, disengagement from learning materials, and overall poor academic performance.

Increased productivity is also a result of personalized learning. Understanding each student's needs allows us to target the areas that demand attention. This implies that content that is already familiar may be omitted, and more time can be spent on topics and places that are proving difficult. By delivering material as students demand it, the learning journey becomes more efficient, allowing instructors and students to engage in areas that require attention. The variation and diversity in such qualities make it extremely difficult to establish a prospective leading solution that meets all demands. Here is when the value of personalized learning becomes clear.

Personalizing training and learning experiences for each student in order to meet their individual requirements, interests, and aptitudes is known as personalized learning. It entails creating curricula and learning activities for each student that are pertinent and interesting based on their interests, advantages, and learning preferences. With the provision of the resources and assistance they require to thrive, personalized learning seeks to assist students in realizing their full potential. This strategy can include leveraging technology to design personalized learning pathways, offering specialized instruction or coaching, and motivating students to take charge of their education by establishing objectives and keeping track of their advancement.

Several venues, such as conventional classrooms, online learning environments, and homeschooling, can use this sort of learning. It has been demonstrated to increase student engagement, motivation, and academic achievement while fostering 21st-century abilities like problem-solving and critical thinking.

- **Student-centred personalized learning**

Students can pick their own method of learning in the classroom using personalized learning. According to their requirements and interests, they choose the learning model. Students actively endeavour to accomplish specified learning objectives, goals, and assignments.

- **The Learning Speed is determined by the students**

Students can choose their own pace thanks to the flexibility of Personalized Learning. Only if they have a firm grasp of the present concept may they decide to move on to subsequent modules. Depending on how complicated the subjects are, students can split their time accordingly.

- **Knowledge-Based Methodology**

Understanding the concepts is the main goal of personalized learning, not getting high exam results. Instructors pay closer attention to what and how kids are learning. Hence, teachers provide opportunities for students to build practical real-life skills rather than merely preparing them for tests and outcomes.

- **Anywhere, Anytime: Learn**

Access to learning and teaching applications is available to both students and teachers both within and outside of the school and classroom. Anyone has access to educational apps thanks to cloud technologies for education.

7.16 CONCLUSION

In conclusion, based on the few data available, several components of personalized learning appear to show potential for enhancing the higher education system in India. To provide causal proof that the approach improves student results, further research is needed. Further study will be required to determine the specifics of which techniques and what combinations are most beneficial for individual pupils because personalized learning is made up of so many interconnected tactics. Early adopters of personalized learning are now constrained by rules that might thwart their efforts, incomplete evidence, and inadequate curricular materials. There is a chance that these issues might lead to early implementations failing as personalized learning techniques expand. This could force the broader idea to be dropped before it can be tried out under more favourable circumstances. Implementers should follow certain guiding principles to identify the facets of personalized learning that are most likely to be successful as a safeguard against these hazards.

REFERENCES

[1] McCarthy, J. (2015). Student-centered learning: It starts with the teacher. *George Lucas Educational Foundation*. https://www.edutopia.org.

[2] Basye, D. (2018). Personalized vs. differentiated vs. individualized learning. Retrieved February 28, 2021, from https://www.iste.org/explore/Educationleadership/Personalized.

[3] Hidden Curriculum. (2014). In S. Abbott (Ed.), the glossary of education reform. http://edglossary.org/hidden-curriculum.

[4] Jenkins, S., Williams, M., Moyer, J., George, M., & Foster, E. (2016). The shifting paradigm of teaching: Personalised learning according to teachers. *Knowledge Works*. Retrieved from www.knowledgeworks.org/sites/default/files/u1/teacher-conditions.pdf.

[5] Patrick, S., Kennedy, K., & Powell, A. (2013). *Mean what you say: Defining and integrating personalized, blended, and competency education*. INACOL. https://www.inacol.org.

[6] Vermont Agency of Education (2017). Personalized learning. Retrieved from http://education.vermont.gov/student-learning/personalized-learning.

[7] Wang, T.I., Tsai, K.H., Lee, M.C. & Chiu, T.K. (2007). Personalized learning objects recommendation based on the semantic-aware discovery and the learner preference pattern. *Educational Technology & Society*. 10 (3), 84–105.

[8] Powell, P. (2019). What is personalized learning? A complete guide to individualized education in the classroom. *Future Readiness*. https://xello.world/en/blog/.

[9] Slocum, N. (2016). What is personalized learning? *Education Domain*. https://aurora-institute.org.

[10] Tanenbaum, C., Le Floch, K. & Boyle, A. (2013). Are personalized learning environments the next wave of K–12 education reform? *American Institutes for Research*. Retrieved from http://www.air.org/sites/default/files/

[11] Laurans, E., Derr, K. & Turco, J. (2013). The state of personalized learning in districts: Research Overview. *The Parthenon Group and Bill & Melinda Gates Foundation*. https://www.learningpersonalized.com.

[12] Wolf, M. (2010). *Innovate to education: System [re] design for personalized learning. A report from the 2010 symposium*. Washington, DC: Software & Information Industry Association. Retrieved from http://siia.net/pli/presentations/PerLearnPaper.pdf.

[13] Yonezawa, S., McClure, L., & Jones, M. (2012). *Personalization in schools*. Washington, DC: Jobs for the Future. Retrieved from http://www.studentsatthecenter.org

[14] Rooney, T., Brown, L., Sommer, B., & Lopez, A. (2017). Beyond reform: Systemic shifts toward personalized learning. Lindsay Unified School District. Contributing researchers: Leslie Pynor, Katie Strom, Eric Haas, and Robert J. Marzano. iNACOL, The International Association for K–12 Online Learning. www.inacol.org.
[15] Tranquillo, J. & Stecker, M. (2016). Using intrinsic and extrinsic motivation in continuing professional education. *Surg Neurol Int. 2016.* 7 (Supply 7): S197–S199. doi:10.4103/2152-7806.179231.
[16] Cherry, K. (2016). Extrinsic vs. intrinsic motivation: What's the difference? http://psychology.about.com.
[17] Great Schools Partnership (2015). Personalised learning. Retrieved February 24, 2018, from The Glossary of Education Reform: https://www.Edglossary.org/personalized-learning.
[18] Rickabaugh, J. (2016). *Tapping the power of personalized learning: A roadmap for school leaders.* Alexandria, VA: ASCD. https://www.teachingchannel.com/k12-hub/blog/personalized-learning-increases-intrinsic-motivation/.
[19] Bray, B. & McClaskey, K. (2017). *How to personalize learning: A practical guide for getting started and going deeper.* Thousand Oaks, CA: Corwin. The Institute for Personalized Learning. (n.d.). Retrieved August 5, 2018, from http://institute4pl.org.
[20] Bray, B. & McClaskey, K. (2013a). A step-by-step guide to personalize learning. Learning and leading with technology. *International Society for Technology in Education,* 40(7), 12–19. Retrieved from http://www.learningandleadingdigital.com/
[21] Bray, B. & McClaskey, K. (2013b). Personalization vs. differentiation vs. individualization report. Retrieved from http://www.slideshare.net/bbray/personalization.
[22] Bray, B. & McClaskey, K. (2013c). International society for technology in education, learning & leading with technology. 12–19. https://www.iste.org.
[23] Miliband, D. (2006). Choice and voice in personalized learning. In *Personalising Education* 21–30. OECD. 10.1787/9789264036604-2-en.
[24] Mohd, C., Shahbodin, F. & Pee, N.C. (2014). Exploring the potential technology in personalized learning environment (PLE). *J. of Applied Science and Agriculture,* 9 (18), 61–65.
[25] Nandigam, D., Tirumala, S.S. & Baghaei, N. (2015). Personalized learning: Current status and potential. *IC3e 2014 -2014 IEEE Conf. e-Learning, e-Management e-Services.* 111–116.
[26] Office of Educational Technology (2017). *Reimagining the role of technology in education: 2017 National Education Technology Plan update. U.S. Department of Education.* Washington, DC: U.S. Department of Education.
[27] Pane, J.F., Steiner, E., Hamilton, M.L. & Pane, J.D. (2017). *Informing progress: Insights on personalized learning implementation and effects.* Santa Monica, CA: RAND Corporation.

Chapter 8

Environment for personalized learning

Jagjit Singh Dhatterwal[1], Kuldeep Singh Kaswan[2], and Sunil Kumar Bharti[3]

[1]Department of Artificial Intelligence & Data Science, Koneru Lakshmaiah Education Foundation, Vaddeswaram, Andhra Pradesh, India

[2]School of Computing Science and Engineering, Galgotias University, Greater Noida, India

[3]Department of Information Technology, Galgotias College of Engineering & Technology, Greater Noida, India

8.1 INTRODUCTION

Many students who perform poorly in science during their secondary education go on to major in non-scientific fields during their post-secondary education, which raises substantial concerns. It is certain that pupils would lack assistance and counsel on how to study and excel in scientific courses despite the hurdles since professors are frustrated with the science learning environment [1]. Moreover, there seems to be no improvisation in handling the scientific classroom, which leads instructors to gloss over challenging ideas. Goodrum also points out that there is a theoretical risk that teaching scientific courses would lead students to see science as overly difficult and, hence, less relevant to students' everyday lives [2]. The teaching and learning process, as well as teacher-student contact in schools, is cited as having a significant impact on student's attitudes toward scientific courses. Goodrum and Rennie both stress that students' "perceptions that, there is little relevance of science courses with their daily life" contribute to the loss of students' enthusiasm for science disciplines. In recent years, internet education has been more integrated into the educational process. The approach used is what sets online education apart from more conventional forms of education [3]. On the other hand, the goals, resources, texts, and schedules are all the same. When content is received over the internet, it is considered online learning because of the structure and process that links students with the resources and information. In most classroom settings, students and teachers seldom interact face-to-face. According to their research, almost 4.8 million students shifted to online programs in 2008.

Web-based learning has become one of the most widely used online applications in the field of education. Web-based education has been proven to improve education in several research investigations. This is because the

DOI: 10.1201/9781003376699-8

Internet has opened exciting new pedagogical possibilities for both students and teachers. The ability to study whenever and anywhere is what makes web-based education so appealing. As was previously noted, the World Wide Web is a highly individualized, worldwide disseminable medium for delivering information. This means that the process of education is no longer restricted by time and location. Students and instructors may be in different places at different times by utilizing this medium to communicate and collaborate. Time and space constraints that formerly defined the classroom are now irrelevant (Bachari et al., 2011).

It is common knowledge that the major benefit of online learning is the opportunity for non-linear engagement. This empowers students to shape their own educational experience. Some pupils may struggle with issues like disorientation, cognitive overload, and lack of self-control if given too much leeway. Consequently, academics are now shifting their focus to discovering how learners of different styles and characteristics utilize web-based learning in order to investigate these issues. It is becoming more necessary to take students' unique learning styles into account when developing and delivering courses (Chatti et al., 2010).

Teachers need to adapt their methods to the unique characteristics of online learning so that students may succeed regardless of their preferred learning style. Therefore, teachers need knowledge about student traits that may influence student behaviour in the classroom. Teachers may tailor their classroom setup to the individual needs of their students in this way [4].

Educators were given a list of factors to consider, including students' age, gender, cognitive style, motivation, and how much previous knowledge they already had in the subject area. Once the learner profile is established, the hypermedia learning process may be tailored to each individual's requirements.

The term "personal learning environment" (PLE) describes a relatively new phenomenon in the modern educational system. The term "Personal Learning Environment" (PLE) refers to a set of technologies that facilitate interaction between learners (or anybody else) and a network of individuals, organizations, and resources. The idea of PLE is relatively new to the field of creating effective online education. In a PLE setting, the learner takes centre stage as opposed to the teacher, classroom, materials, and technology. Likewise, PLE has been useful in boosting the efficiency of education. PLE, as described by Atwell, is a personalized setting. Each person is accountable for his or her own education. They must also take a more active role in managing the learning process and in the content's ownership. In general, a personalized learning strategy may provide a fresh way to motivate pupils to study and may even be necessary to satisfy the educational demands of the future [5].

Science Two is the focus of this investigation. Specifications for an Integrated High School Curriculum. The Malaysian Ministry of Education's framework for Science Form 2 is used. As a result of this work, a novel method of education called the Personalized Learning Environment will be proposed. (PLE). In PLE, the learner takes centre stage as opposed to the teacher, classroom, materials, and technology. PLE may play a vital function in enhancing the efficiency of learning. A personal learning environment (PLE) is one of the resources available to a student in a network of people, services, and resources [6].

8.2 ABOUT PERSONALIZED LEARNING STRATEGY

In the twenty-first century, children's particular strengths, interests, and aptitudes are recognized and cultivated via a personalized learning strategy. Likewise, PLE has been useful in boosting the efficiency of education [7]. Indeed, conventional learning based on a "one size fits all" approach tends to support just one educational model since a teacher in a normal classroom environment typically needs to deal with numerous pupils at once. To ensure that students are actively involved in and taking responsibility for their own education, several scholars have argued that consideration should be given to the unique characteristics of each learner [8]. Learners always have power over what they learn, but they may not always have a say in what is taught. One of the most important aspects of customization is the learning experience. We utilize Google, social media, online forums, and wikis to collaborate on assignments and solve problems. There is much academic learning that occurs outside of schools. A paradigm shift may be seen in the concept of PLE, which proposes a learning environment in which users have access to a steady stream of information and guidance from a wide variety of sources. Learning takes place in a wide variety of settings, making this tool crucial for facilitating this process [9].

Learners vary in numerous ways, including their learning styles, orientations, learning rates, cognitive styles, level of intellect, number of innate abilities, and other facets of their potential.

This research took into account three distinct learning styles: aural, visual, and kinetic. Students learn best via visual means, such as pictures, images, and spatial comprehension. Observe and absorb. Take notes, listen carefully, and review your notes often. They are able to participate in class discussions and lectures without difficulty. Summarize the material and speak it aloud after reading [10].

Students like auditory methods such as music and sound effects. Auditory learning. Take notes often and swiftly evaluate visuals. They train their minds to think more clearly by mentally visualizing certain words or ideas. They also make use of the following tools:

Students who learn best using kinesthetic methods are those who learn best by doing. Feel your way to understanding. Repeated writing will help you retain the information. We save the scratch paper. Taking and retaining notes from lectures is crucial. Create some study guides [11].

Adopting apps, matching educational pursuits, and incorporating technological components into teaching all contribute to a successful PLE deployment. As a result, classroom setup should cater to individual students' preferences, strengths, and weaknesses.

8.3 METHODOLOGY

The methodology used to develop a courseware is ADDIE (Analyze, Design, Develop, Implement, and Evaluate) model. ADDIE is a systematic or step-by-step model used for product development. Each phase ensures development efforts stay on track, time, and target [12].

8.3.1 The research process

During this stage, we analyze the data to determine the root causes of the students' difficulties with the Science curriculum. The evaluation needs of students were then taken into account when we set our goals and built the PLE environment and approach. In order to confirm the need, we will undertake surveys, focused group interviews, and a review of relevant literature [13]

8.3.2 The planning stage

Create a test version of your suggested model that emphasizes learning. The anticipated results from the aforementioned study may then be used to inform the design of the interactive prototype's content, media, and interaction.

8.3.3 In development

PLE components are integrated throughout development. Now that the storyboard has been made, development can begin, and eventually, an Alpha version of the product will be the result. Using the storyboard as a blueprint, the team will first work on the user interface and then add content. We will construct each module and test its features to make sure they all operate [14].

8.3.4 The stage of implementation

In this stage, the prototype is tested to ensure it works properly. Each component and module will be combined into one fully functional whole. It is crucial to run a demo of the application with a select audience in order to get valuable input that can be included in future iterations of the program.

8.3.5 The assessment step

During this stage, we identify the processes that may be enhanced in order to provide better study outcomes. User comments are gathered as part of the evaluation process. Testers' responses to a questionnaire on the program's user interface, PLE approach, and content will be utilized to improve the product.

8.4 PERSONALIZED LEARNING ENVIRONMENT IN EDUCATION

To better meet the individual requirements, traits, and preferences of each student, educational institutions are increasingly turning to personalization technology. Learners who take charge of their own education and progress use PLEs, which consist of a collection of tools, communities, and services. Personalized learning proponents argue that the notion should be modified for each student, rather than the other way around, in contrast to the standard method of education [15].

By focusing on each student's unique set of circumstances, a personalized approach to education may foster a close, one-on-one interaction between the classroom and the classroom's students. Learner modelling is the primary technique for tailoring the system's interaction with each individual student in a customized learning environment. Generating a learner model entails hypothesizing the learner's intentions, plans, preferences, attitudes, knowledge, and beliefs based on the information obtained via interaction [16].

According to studies, pupils who study in a more individualized setting improve both academically and socially. These pupils are becoming more independent, creative, and resourceful. Personalized learning environments (PLEs) include a number of characteristics, as described by Clements and Douglas (2008) in their paper Personalized Learning and Innovation in Education. This includes:-

- It gets students involved in the learning process, which in turn makes them more responsible for their own success. Students are learning to produce content rather than just consuming it.
- Students take responsibility for their learning.
- Gives kids the independence they want
- It is a link to the actual world
- It encourages kids to think outside the box
- Promotes in-depth thinking, learning, and comprehension
- Allows for the free exchange of ideas
- Fosters a relationship based on trust and cooperation between instructors and their students.

8.5 WHY PERSONALIZED LEARNING ENVIRONMENT?

The introduction of online courses has revolutionized the educational system. It is unfortunate that many supposed educational websites do not use sound pedagogical practices. How can we best target each user as they increasingly turn to online education? How can we foster more initiative, autonomy, and student responsibility? How can I tailor a website's layout to each user? How can I tailor my message to a wide range of personality types by using a variety of settings? How do various students use the online curriculum? In what ways do our consumers vary from one another? The answers to these sorts of inquiries are crucial when creating an instructional website [17].

Research on individual differences and needs has emerged as a central concern during the last decade as a means of addressing these concerns. The significance of adjusting for different users while creating online guides is a good indicator of this. The customization problem emerges when we consider how best to accommodate students' unique needs by allowing them to tailor their classroom experience to their own unique set of experiences and backgrounds.

The absence of individualization has been identified as a major flaw in virtual classrooms. Therefore, personalized education is one of the most pressing concerns in the field of education today. According to proponents of the customized learning approach, this style of instruction puts the learner, or student, at the centre of the educational process. Students are more invested in their education, so they study at their own speed and in their own way, are more motivated, and achieve more uniformity in their knowledge. Otherwise, students will take advantage of self-paced learning settings, where they may choose their own tempo, get the information they need, choose from a variety of teaching methods, and manage their time effectively. As a result, we need further studies examining the connections between student background information and the best ways to provide course materials online. This research is necessary because it will help us understand what features of the online learning delivery format lead to better student happiness and performance.

8.6 INDIVIDUAL DIFFERENCES IN PERSONALIZED LEARNING ENVIRONMENT

Knowing your students is essential for creating a productive learning atmosphere. The extent to which individual variations affect the learning process has been the subject of much study. Differences in cognitive styles and background information are examined in more detail below.

8.7 COGNITIVE STYLE

Each person has their own unique way of thinking and processing information, and this is called their "cognitive style." Cognitive style is said to influence how one perceives the world, processes information, and ultimately comes to conclusions and choices. A person's cognitive style may be defined as the way in which they naturally and most effectively acquire and process new information and solve problems.

There have been many studies done on the cognitive style factor of field dependency vs. field independence. The reason for this is that the ways in which people learn best – through salient signals and field arrangement – depend on the cognitive style they employ. Most empirical research focuses on the following questions:

To what extent (a) various groups of students with different cognitive styles prefer utilizing different sorts of navigation tactics, and (b) whether cognitive styles will substantially impact learners' success inside web-based training. Field independence was used to define people who are self-reliant, analytical, and focused inward. Field-reliant people, on the other hand, are more likely to operate in teams, take direction from others, be swayed by prominent factors, and accept ideas at face value. Due to the differences, field-independent learners have been shown to outperform field-dependent learners in a variety of traditional and online learning contexts [18].

8.8 PRIOR KNOWLEDGE

One factor contributing to individual variances is a person's level of prior knowledge. An individual's prior knowledge incorporates a comprehension of their past experiences. Numerous research studies have shown the importance of previous knowledge in online training. It is well acknowledged that issues of disorientation and the need for extra help are significant challenges in online education. It has been shown via studies that individuals with less knowledge and expertise would have greater difficulty finding their way around in a web-based tutorial. The fact that they are not yet acquainted with the new material means that they have no frame of reference within which to place it. Knowledgeable users, on the other hand, will not have any trouble building new knowledge and making connections to existing information.

8.9 PRELIMINARY ANALYSIS

Ninety second graders at SMK Malim in Melaka, Malaysia, are polled to determine what aspect of Science is the most challenging for them.

The results of this preliminary research on the most challenging areas of Science Form 2 are shown. The second-year science professors are interviewed by the researcher to determine the most challenging course material. The complexity and breadth of nutrition's various subfields make it the toughest. Food groups, the value of a well-rounded diet, the digestive process, digested food absorption, urine and feces reabsorption, and good eating habits are all components of the broader field of nutrition.

To evaluate the survey data, we use SPSS (Statistical Package for the Social Sciences). (SPSS). Students have been interviewed sparingly for their thoughts on the issue of nutrition. According to the comments, they should keep in mind the information covered in class. This subject was likewise dull to them. As students have not been introduced to ways to enhance the learning processes in this area, their boredom has grown. Topic 2 is Nutrition in Science, and this table displays the relative frequency of its occurrence. Nutrition is the most challenging subject, as agreed upon by 40%, and highly agreed upon by 38.9%. The next most popular answer, "easy," was given by 12.2% of respondents, followed by Nutrition (chosen by 3.3%).

The data demonstrates the investigation of how much kids rely on the internet for schoolwork. Statistics reveal that kids require internet access for their education with 40% strongly agreeing and 38.9% agreeing. Only 2.2% disagree and 1.1% strongly disagree that they used the internet for schoolwork. They place a high importance on practical resources that aid in task organization, time management, the simplification of complex jobs, and, of course, entertainment.

This study will provide a framework for recognizing and evaluating students' preferred learning styles, including the three most common ones: visual, auditory, and kinesthetic. Students will be evaluated based on their actions and progress. The goal of this project is to create a model and framework that puts an emphasis on the mental abilities necessary for PLE.

8.10 CONCLUSIONS

It is likely not unexpected that the study's preliminary analysis suggests that computers are being utilized in classrooms in ways like those of previous classroom uses of technology. If we are going to make the most of the potentials for PLE discussed in the literature, then we need to put in more effort to ensure that teachers receive successful instruction to understand what those potentials are, and that policymakers ensure that enough people have access to the Internet to make that kind of movement necessary.

REFERENCES

[1] Alkhasawneh, I. M., Mrayyan, M. T., Docherty, C., Alashram, S. A., & Yousef, H. Y. (2008). Problem-based learning (PBL): Assessing students' learning preferences using vark. *Nurse Education Today, 28*(5), 572–579. 10.1016/j.nedt.2007.09.012.

[2] Aviram A., Ronen Y., Somekh S., Winer A., & Sarid A. (2008). Self-Regulated Personalized Learning (SRPL): Developing iClass's pedagogical model. *eLearning Papers*, 9, 1–17.

[3] Attwell, G. (2007). Personal learning environments - The future of Learning? *eLearning Papers*, 2(1). http://digtechitalia.pbworks.com/w/file/fetch/88358195/Atwell%202007.pdf

[4] Kaswan, K. S., Dhatterwal, J. S., Sharma, H., & Sood, K. (2022). Big Data in Insurance Innovation. Sood, K., Dhanaraj, R. K., Balusamy, B., Grima, S. and Uma Maheshwari, R. (Ed.) *Big Data: A Game Changer for Insurance Industry (Emerald Studies in Finance, Insurance, and Risk Management)*, Emerald Publishing Limited, Bingley, pp. 117–136. 10.1108/978-1-80262-605-620221008.

[5] Elliott, C. (2010). We are not alone: The power of Personal Learning Networks. *Synergy*, 7(1), 47–50.

[6] Ertl, B., Ebner, K., & Kikis-Papadakis, K. (2010). Evaluation of eLearning. *International Journal of Knowledge Society Research (IJKSR)*, 1(3), 31–43, 2010.

[7] Dhatterwal, J. S., Kaswan, K. S., & Pandey, A. (2022). Implementation and Deployment of 5G-Drone Setups. In *The Internet of Drones* (pp. 47–64). Apple Academic Press.

[8] Gagné, R., & Briggs, L. J. (1974). *Principles of Instructional Design.* http://ci.nii.ac.jp/ncid/BA04482406

[9] Gu, X., & Li, X. (2009). *A Conceptual Model of Personal Learning Environment Based On Shanghai Lifelong Learning System. Proceedings of the 17th International Conference on Computers in Education [CDROM]*. Hong Kong: Asia-Pacific Society for Computers in Education, 885.

[10] Kaswan, K. S., Dhatterwal, J. S., Kumar, S., & Lal, S. (2022). Cybersecurity Law-based Insurance Market. In *Big Data: A Game Changer for Insurance Industry* (pp. 303–321). Emerald Publishing Limited.

[11] Goodrum, D., Rennie, L. J., & Hackling, M. W. (2000). *The Status and Quality of Teaching and Learning of Science in Australian Schools: A Research Report.*

[12] Minocha, S., Schroeder, A., & Schneider, C. (2011). Role of the educator in social software initiatives in further and higher education: A conceptualisation and research agenda. *British Journal of Educational Technology*, 42(6), 889–903. 10.1111/j.1467-8535.2010.01131.x

[13] Retalis, S., Paraskeva, F., Tzanavari, A., & Garzotto, F. (2004). Learning Styles and Instructional Design as Inputs for Adaptive Educational Hypermedia Material Design. *Paper presented at the "Information and Communication Technologies in Education" - Fourth Hellenic Conference with International Participation*, Athens, Greece.

[14] Samah, N. A., Yahaya, N., & Ali, M. B. (2011). Individual differences in online personalized learning environment. *Educational Research Review*, 6(7), 516–521. 10.5897/err.9000199

[15] Downes, S. (2006). Learning Networks and Connective Knowledge. *Collective Intelligence and Elearning*, 20, 1–26.

[16] Kaswan, K. S., Dhatterwal, J. S., & Kumar, A. (2023). *Swarm Intelligence: An Approach from Natural to Artificial.* John Wiley & Sons.
[17] Speering, W., & Rennie, L. J. (1996). Students' perceptions about science: The impact of transition from primary to secondary school. *Research in Science Education, 26*(3), 283–298. 10.1007/bf02356940.
[18] Trinidad, S. (2003). Working with Technology-Rich Learning Environments: Strategies for Success. *World Scientific eBooks*, 97–113. 10.1142/9789812564412_0005

Chapter 9

AI in personalized learning

Kuldeep Singh Kaswan[1], Jagjit Singh Dhatterwal[2], and Rudra Pratap Ojha[3]

[1]School of Computing Science and Engineering, Galgotias University, Greater Noida, India
[2]Department of Artificial Intelligence & Data Science, Koneru Lakshmaiah Education Foundation, Vaddeswaram, Andhra Pradesh, India
[3]Department of Computer Science & Engineering, GL Bajaj Institute of Technology and Management, Greater Noida, India

9.1 INTRODUCTION

The rise of digitalization as a driver of economic expansion and improvement in living standards is well recognized. Changes in global, creative, inclusive, and sustainable growth across industries are influenced by digitalization [1]. Computational science has flourished as a result of the availability of massive datasets and the rising importance of digital technology, which in turn has stimulated the expansion of many sectors and permitted a swift transition to e-science on a national scale. Computer literacy has emerged as a key competency driving technological advancements in the business world, thanks to the rise of smart computing [2]. Additionally, it has been acknowledged as a factor in the expansion of businesses. The broad accessibility of low-cost computing power and vast volumes of data has contributed to recent advancements in machine learning. The widespread dissemination and accessibility of digital technology have sparked a renewed interest in artificial neural networks in both industry and academia. Among the various applications of neural AI and ML are real-time language processing, translation, image analysis, and driverless vehicles. Examples include synthetic artwork, service robots, automated fraud identification systems, and automated procedure systems for administration.

AI is revolutionizing many sectors of economic life and communication, while also bringing virtual reality to the forefront. Learning, cognition, and the progress of society have all been impacted by AI's introduction into the classroom. The biggest problem is that schools still do not believe in the usefulness of technology for learning, which has delayed the introduction of AI into classrooms for a long time [3]. However, digital changes have occurred in the fields of applied sciences, industry, finance, and health. As a result of the digital revolution of education, secondary and higher education institutions all over the globe have become major adopters of technology. It

DOI: 10.1201/9781003376699-9

has also stoked studies into the practical use of AIEd (artificial intelligence in education) in contemporary classrooms.

To improve education, technological advances are being used in the classroom, however, many developed countries still face challenges in doing so. Due to the interplay of numerous factors, the successful application of intelligent machines (AI) to the objective of supporting society's long-term progress needs the collaboration of a wide range of organizations. Many educational procedures, such as determining whether or not they are effective and whether or not they make sense, are extremely subjective, and not all public educational policies are uniform around the globe. Although artificial intelligence (AI) products have the potential to enhance students' opportunities for higher education, interactive instructional spaces, and intelligent teaching assistants, putting these ideas into practice is still easier said than done. It is, therefore, relevant to investigate how intelligent technology might be used to best promote personalized education in the classroom [4].

The focus of this research is on how AIEd may be utilized to tailor education to the specific needs of each student. To do so, the following objectives are set:

> Describe the digitization of society and the potential social and ethical risks posed by AI; Explain the meaning of a certain AI-related word, key ideas, and subfields; Investigate the potential of AIEd and look at other ways of learning and growing in your field; Study real-world examples of using AIEd to learn more; Analyze the ways in which AIEd may facilitate personalized education [5].

Students now have to study at a much faster rate than their predecessors because of the increasing relevance of AI. However, there is broad public concern that the deployment of AI to tackle social and economic concerns may have unexpected repercussions. When applied to learning, AIEd opens up exciting new horizons. In the relatively short history of AIEd, there have been a number of key moments that may be categorized into three overarching paradigms of how AI is applied to difficulties in education and learning. In the first approach, students are seen as passive consumers of educational services while conceptual models of knowledge are represented and cognitive learning is directed by means of AIEd technology. In the second model, students work in tandem with AI and use AI-enhanced pedagogical materials. In the third framework, students use AIEd tools to become more involved in their own education. With the aim of creating iteratively upgraded customized, learner-centred, data-driven training that emphasizes review and criticism from students to artificially intelligent machines, The application of computational intelligence in education (AIEd) is progressing towards individualizing learning experiences for students. While individualized instruction is not new, advances in AI and big data analysis are making it a lot more practical for use in educational institutions.

The online classroom provides the flexibility to cater to different learning styles. However, empirical studies show that the mere fact of customization does not guarantee improved educational outcomes. Attempts at personalization in the classroom may fail if they are based on static modelling or if they make adjustments based on student traits that are only loosely related to the method of instruction and instruction. (Such as learning styles). However, dynamic modelling can assess and adapt to important learner traits, resulting in learning performance that rivals that of human effort.

Intelligent Learning Systems may be very useful for teachers who focus on online or hybrid instruction. (ITS). The widespread availability of digital resources has facilitated the dramatic shift away from in-person classrooms and towards distance education. With its capacity to enable customizable groupings based on student data, promote communication among virtual team members, and provide reports on group conversations., AIEd has the potential to promote learning through collaboration. Artificial school systems are predicated on the principle that each student should have their own unique educational experience tailored to their unique set of characteristics and past knowledge. In contrast, data-driven AI systems excel in real-time processing of massive, complex data streams. User interfaces (UIs) that capture pupil action in real time and save historical data to be utilized in building a student's profile are necessary for the coming generations of intelligent learning systems. This concept is sometimes expressed as "no AI without UI." As a result, significant resources will be devoted to integrating various technologies for sensing and graphical user interfaces into academic settings. They will make it easier to get data on students' habits from outside sources like social media and gaming sites [6].

Artificial intelligence (AI) can help teachers create lessons that are customized for each student. The use of technology to address educational challenges and promote new pedagogical approaches is brought to the forefront of people's minds. The fast development of AI has made it feasible to create unique educational plans for each student. Some applications of artificial intelligence that attempt to emulate human performance include machine translation, voice recognition, computer vision, picture recognition, recognizing texts, theorem demonstrating, algorithmic learning, intelligent learning with adaptation, and robotics [7].

In order to examine students' knowledge (and their learning engagement techniques) and reconstruct effective individualized learning pathways and develop curricular supporting strategies, learning analytics now makes use of AI as one of its tools. Learning data analytics algorithms are widely used for decision assistance and tailored instruction. Predictive analytics is used by systems for decision support to provide predictions based on graphical representations of data. AIEd may be used by teachers to analyze student assessment results. Personalized learning systems might evaluate a student's progress towards mastery of course material and adjust their education appropriately. Students' interests and needs may be better met and learning results improved via the use of personalization strategies. Artificial intelligence (AI) and immersive technology have the potential to make today's classrooms more engaging and inspiring.

Information systems link, govern, and promote the professional and academic communities. New studies on the benefits of technological progress and the development of innovative educational methods are made possible by the progress of AI and other forms of machine learning. The fields of artificial intelligence (AI) and robotics have attracted increasing attention in the past couple of decades. The importance of digital materials in today's workplaces and classrooms is undeniable. The extensive use of digital technologies will have significant effects on both teachers and pupils. Therefore, universities have an obligation to equip their students with the knowledge and abilities they will need to succeed in the interconnected economic system of the 21st millennium and to aid in the process of social adaptation. Educators can do more with AI, and the gap between instructional technology creation and execution in the classroom is narrowed. The advancement of both education and technology is mutually beneficial and should be encouraged simultaneously. Educators and AI professionals must collaborate efficiently since educational technology cannot exist alone. Students' employability is tied to their capacity to acquire marketable skills, which in turn depends on their ability to communicate effectively in the classroom [8].

9.2 BASIC TERMS OF ARTIFICIAL INTELLIGENCE

Understanding how robots can learn as well as logic like humans is a broad field of study known as artificial intelligence [9]. There are both elementary rules and more sophisticated neural networks present. The field of artificial intelligence known as machine learning (ML) focuses on teaching computers to learn and make predictions on their own. Machine learning that makes use of artificial intelligence (AI) is known as deep learning (DL) or deep neural networks (DNN). (Figure 9.1). Manufacturing companies that rely heavily on data commonly employ and combine the techniques of cognitive ability, intelligence, machine learning, and extensive intelligence to create AI-based products and services. Human-centred machine learning (HCML) has emerged as a result of the growing partnership between humans and AI [10,11].

Algorithms are the rules and instructions that computers follow to solve problems and reach their objectives. Algorithms are a kind of instruction that AI and ML systems need in order to perform. Neural networks analyze large amounts of data using complex algorithms to do jobs that were previously only feasible for humans. Machine learning algorithms mine

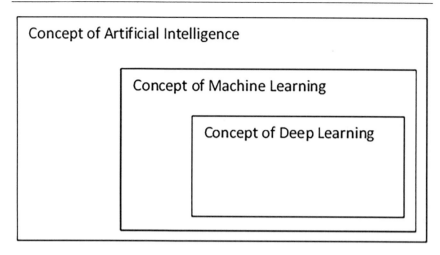

Figure 9.1 Artificial intelligence concepts [12].

information for patterns that may be used to make forecasts about the future. The quality of the information and human expertise utilized to train machine learning is crucial to the success of the process. Machine learning techniques may be found in a wide variety of fields, including the categorization of information, data clustering, regression study, feature design, reduced dimensionality, rule-based learning, and reinforcement learning to name a few. It is not always a breeze to zero down on the best learning algorithm for a given challenge. Various learning algorithms may have various ultimate goals. The results of various learning algorithms on the same sample of data may vary dramatically depending on the data and the approach applied. Computer algorithms that might potentially "learn" from their mistakes by analyzing massive amounts of data are referred to as "machine learning." One of the most crucial aspects of learning is data, which presents a problem for the field of machine learning. Machine learning takes time since it uses series data to assess and explore possible outcomes. Mining data, creating new data, cleaning existing data, analyzing data, and visualizing data are the primary tasks.

The field of machine learning known as "deep learning" makes use of artificial neural networks. When compared to shallow machine learning models and more conventional methods of data analysis, deep learning models perform better. However, it is possible that the widespread use of machine learning has its roots in agricultural instruction. Institutions in many European nations, including the UK, Greece, and Italy, have shown how machine learning may be used to enhance the broad field of agri-technologies by making greater use of existing data. Studies using machine learning to distinguish between ill and healthy wheat crops have also been conducted. By combining hyper spectral absorption information with a self-organizing classification based on a hierarchical structure, this breakthrough has enabled the development of a novel approach for identifying winter wheat with and without yellow rust disease. This has real-world relevance because it paves the way for precise dosing of fertilizers and fungicides on individual plants [13].

9.3 ACQUIRING KNOWLEDGE THROUGH PERSONALIZED LEARNING [14]

The broad field of research known as artificial intelligence (AI) aims to one day replace humans in decision-making roles. One's own set of assumptions, prejudices, opinions, and cultural norms, among other things, are brought to any new knowledge. Acquiring new information, perspectives, and skills is a uniquely personalized endeavour. The typical classroom has the major issue of assuming that all students have the same level of background understanding, instructional style, and end objectives. The standard curriculum guides students towards a certain route to academic achievement [15]. Not every kid benefits from following a set curriculum. Individualized learning plans are created for each student with consideration given to their needs, interests, learning styles, and prior knowledge.

When using student-centred tactics, it is crucial to take into account the individuality of each teacher. Teachers in customized learning settings have less of a "commanding presence" and more of a "coordinator" or "mentor" role than they do in traditional classrooms. IT education for teachers should also emphasize skill acquisition using the tools that will be used in the digitization of STEM topics.

Individualized instruction that takes into account each student's background, interests, and long-term goals is known as "personalized learning." This affords students the chance to develop personally and academically by engaging with a wide range of disciplines. Teachers in a PL classroom should encourage students to take an active role in their education by assisting them in identifying and pursuing goals that are of most interest to them. When used effectively, individualized learning systems and approaches may stimulate students' enthusiasm for learning and boost them to new academic heights.

Both the student's individual growth and their educational setting are given top priority in a personalized learning setting. Services, educational resources, and apps in a tailored learning space are developed with each student specifically in mind. Web 3.0 and Web 2.0 technologies provide for more affordable training, more engaging user experiences, and more individualized profiles. An adaptive system may be set up to review lessons as often as necessary to ensure that students grasp the concepts.

The application of AI in education has the potential to benefit students across the board in terms of age, ability level, and socioeconomic status. Artificial intelligence has helped customize lessons in significant ways.

Artificial intelligence (AI) can assist teachers design individualized lesson plans based on each student's unique set of skills, interests, and challenges [16]. Social and emotional development might benefit from the use of advanced analytics and machine learning. Educators may now review both qualitative and quantitative data because of the wealth of modern tools available to them. Educators' use of AI, data, analytics, and machine learning has the potential to enhance the online learning environment for their students. This method is useful for guaranteeing the efficiency and usefulness of distance education.

The advent of machine learning, artificial intelligence, and big data analytics has opened up exciting new doors for delivering individualized education. One of the most recent developments, personalized learning allows teachers to adapt their lessons to each student's specific needs. The new educational model emphasizes the uniqueness of each learner by tailoring instruction to their specific needs. Lifelong education is a great way to boost your resume and your career prospects. Increasingly, schools are turning to machine learning systems and algorithms to provide personalized learning environments, automated assessment tools, face recognition software, interactive virtual assistants (SNS), and data mining and forecasting programs [17]. Due to its usefulness in assisting both educators and students in identifying and acquiring optimal learning resources, intelligent learning systems have quickly become one of the most popular and extensively used AI applications. A computer-based ITS might potentially instruct students and provide them with real-time, individualized feedback.

In the context of the recent coronavirus epidemic, artificial intelligence has been extensively deployed to speed up education. These changes in schooling had an especially large impact on the field of medicine. Evidence from the University of the West Indies suggests that the COVID-19 epidemic has created a number of challenges for premedical training. These obstacles include the reduction or elimination of hands-on/lab sessions, the discontinuation of cadaver dissections, and the termination of in-person instruction. Conversely, it has spawned a wide variety of novel enterprises, the likes of which would not exist in a world without AI. Therefore, because of its robust potential for virtual simulation, artificial intelligence was shown to be a reliable instrument for adaptive learning [18].

The purpose of AIEd is to provide students with a personalized learning platform that can adjust the pace, difficulty, and grading scale based on their individual requirements. Several processes may be automated, and the student's development can be tracked with the use of artificial intelligence. By taking into account both the classroom atmosphere and the students' achievement, AI aids instructors in implementing the most successful instructional tactics. The AIEd provides students with engaging new options for online learning and personalized, classroom-wide metacognitive clues. Students are more interested and do more work while using AIEd.

Advantages of AIEd for teachers include its potential utility in monitoring student progress, creating individualized courses, assessing student learning, and analyzing data. Automating evaluations, digital asset categorization, and scheduling using AIEd may lead to the quickest return on investment. If teachers have access to AIEd, they may be able to spend less time on administrative tasks and more time with students. Teachers are adapting their methods and curricula to better prepare their students with the future-ready abilities that will be in high demand. Analytics backed by AI is essential for tracking changes in curriculum-relevant subjects. Integrating AI with other IT initiatives, such as connected devices and a managed IoT network, has the possibility of helping greatly improve students' educational prospects. This is according to a 2021 study. With the use of expertise systems, sensory input, and a visual, comprehensive approach to learning, technology, an AI learning environment has the potential to significantly enhance LMS for both instructors and students. Results in the classroom might be enhanced by adopting new teaching strategies and using new technologies. When students get personalized instruction, they are better equipped to take responsibility for their learning and use what they have learned in the real world.

Artificial intelligence (AI) technologies may help improve educational settings. It is conceivable for educational policy and practice to include several worldviews and pedagogical approaches. Students could learn more via creative teaching methods. That is what the data suggests. Personalized learning is now a realistic option because of technological advancements in the classroom and ground-breaking pedagogical techniques. Differentiated education may be delivered in a variety of ways using AIEd technology. The use of cutting-edge AIEd in the classroom has been shown to have positive effects on students' motivation, engagement, and learning. With the assistance of these novel methods, today's students will be better prepared to meet the challenges of tomorrow. Children go up to the plate when they are presented with an opportunity to stretch their talents. For instance, scientists from Nanyang Technological University in Singapore have investigated the usefulness of VR for instructing future doctors. Medical students may benefit from using virtual reality to help them understand difficult ideas. Another study focused on the potential of digital gaming to help students develop business acumen. Students were able to learn more about business with the help of this one-of-a-kind online game [19].

The development of deep and network learning and the creation of tools for evaluating large volumes of data may be traced back to shifts in NLP. Artificial intelligence (AI) can now assess a person's mental health by recognizing negative emotions such as stress, anxiety, and sadness. The Intelligent Computer-Assisted Language Learning system was created as a result of research into how people acquire a second language. (ICALL system). The system consists of a variety of intelligent technologies, including a virtual reality environment that uses adaptive learning algorithms to facilitate collaboration education. Individualized course materials, machine translation applications, chatbots, and more are also a component of the language-learning ecosystem powered by artificial intelligence. There are both elementary rules and more sophisticated neural networks present. The field of artificial intelligence known as machine learning (ML) focuses on teaching computers to learn and make predictions on their own. Machine learning that makes use of artificial intelligence (AI) is known as deep learning (DL) or deep neural networks (DNN) (Figure 9.1). Industries that rely heavily on data commonly employ and combine the application of artificial intelligence, machine learning, and deep learning to create AI-based products and services. Human-centred machine learning (HCML) has emerged as a result of the growing partnership between humans and AI [20].

Algorithms are rules and guidelines that computers follow to solve problems and achieve objectives. Algorithms are a kind of instruction that AI and ML systems need in order to perform. Machine learning's end objective is to eliminate the need for people to do certain activities by using large datasets and intricate algorithms. Artificial intelligence algorithms are designed to mine data for predictable patterns and models. For algorithmic learning to be effective, training data and information must be of high quality. Some common algorithms used in machine learning include those for classification, regression, clustering, data feature design, decreasing dimensionality, rule-of-association learning, and learning through reinforcement. It is not always a breeze to zero down on the best learning algorithm for a given challenge. Various learning algorithms may have various ultimate goals. The results of various learning algorithms on the same sample of data may vary dramatically depending on the data and the approach applied [21].

Computer algorithms that might potentially "learn" from their mistakes by analyzing massive amounts of data are referred to as "machine learning." One of the most crucial aspects of learning is data, which presents a problem for the field of machine learning. Machine learning takes time since it uses series data to assess and explore possible outcomes. Mining data, creating new data, cleaning existing data, analyzing data, and visualizing data are the primary tasks. The field of machine learning known as "deep learning" makes use of artificial neural networks. When compared to shallow machine learning models and more conventional methods of data analysis, deep learning models perform better. However, it is possible that the widespread use of machine learning has its roots in agricultural instruction. Institutions in many European nations, including the UK, Greece, and Italy, have shown how machine learning may be used to enhance the broad field of agri-technologies by making better use of existing data. Studies using machine learning to distinguish between ill and healthy wheat crops have also been conducted. Specifically, this breakthrough has allowed researchers to create a novel approach to identifying winter wheat with and without yellow rust disease by using hyperspectral reflectance data and a hierarchical self-organizing classifier. This has real-world relevance because it paves the way for precise dosing of fertilizers and fungicides on individual plants.

However, the theoretical underpinnings for choosing how to use AI as a digital tool in the classroom vary widely from one location or institution to the next. Personalized student instruction, which is presently adopting a more conventional shape as a solution to the world's most pressing concerns, has not been properly investigated in the context of numerous works describing the foundations of introducing AI into the classroom.

9.4 AI IMPLEMENTATION IN EDUCATION

AI implementation in education refers to the integration of artificial intelligence technologies and applications to enhance teaching, learning, and administrative processes in educational settings. AI has the potential to transform various aspects of education by providing personalized learning experiences, automating administrative tasks, and facilitating data-driven decision-making. Here are some areas where AI is being implemented in education:

Personalized Learning: AI can adapt instructional content and delivery to meet the individual needs and preferences of learners. Intelligent tutoring systems use AI algorithms to provide personalized feedback, suggest appropriate learning resources, and adapt the pace and difficulty level of the material based on the student's progress and performance [22].

Intelligent Assessment: AI-powered assessment tools can automate the grading and feedback process, saving teachers time and providing immediate feedback to students. Machine learning algorithms can analyze student responses, detect patterns, and generate insights into individual and group performance, helping educators identify areas of improvement.

Virtual Assistants and Chatbots: Virtual assistants and chatbots can be used to provide instant support to students, answering their questions, guiding them through learning materials, and assisting with administrative tasks. These AI-driven chatbots can be integrated into learning management systems or educational websites to provide on-demand assistance and support.

Adaptive Learning Platforms: Adaptive learning platforms use AI algorithms to analyze student data and provide customized learning paths. By continuously assessing a student's performance and adjusting the learning content and activities accordingly, these platforms can optimize learning experiences and promote individual mastery of concepts.

Administrative Efficiency: AI can streamline administrative tasks in educational institutions. For example, AI-powered systems can automate administrative processes like enrollment management, scheduling, and student record keeping. This automation reduces administrative burden, enhances efficiency, and allows educators to focus more on teaching and student support.

Data Analytics and Insights: AI algorithms can analyze large volumes of educational data to identify patterns and trends. This data-driven approach enables educators and administrators to gain insights into student performance, identify at-risk students, and make informed decisions about curriculum design, instructional strategies, and resource allocation.

Language Learning and Translation: AI-powered language learning platforms can provide personalized language instruction, pronunciation feedback, and language practice exercises. AI-based translation tools can assist students and educators in overcoming language barriers, facilitating communication, and collaborating in multilingual educational settings.

Intelligent Content Creation: AI technologies can assist in the creation of educational content. For instance, AI can generate interactive simulations, virtual reality experiences, and multimedia materials to enhance the learning experience and engage students in immersive and interactive activities.

9.5 ROLE IN THE DIGITALIZATION OF EDUCATION

As an AI language model, I can provide information on the role of digitalization in education. Digitalization has had a profound impact on education, transforming traditional teaching and learning methods. Here are several key roles digitalization plays in the field of education:

Access to Information: Digitalization has made vast amounts of information readily available to students and educators. Online resources, e-books, and digital libraries provide instant access to a wide range of educational materials, allowing students to explore subjects beyond their textbooks and learn at their own pace.

Enhanced Learning Experience: Digital tools, such as interactive multimedia presentations, simulations, and virtual reality, enrich the learning experience by making it more engaging and interactive. These tools help students visualize complex concepts, conduct virtual experiments, and participate in immersive learning activities [23].

Personalized Learning: Digitalization enables personalized learning experiences tailored to each student's needs and abilities. Adaptive learning platforms use data analytics to assess students' progress and provide personalized recommendations, allowing educators to offer targeted support and interventions [24].

Collaboration and Communication: Digital platforms facilitate collaboration and communication among students and educators. Online discussion forums, collaborative document editing, and video conferencing tools enable remote teamwork, peer-to-peer learning, and global collaboration.

Flexibility and Accessibility: Digitalization offers flexible learning options, breaking down barriers of time and location. Online courses, distance learning programs, and virtual classrooms enable learners to access education from anywhere, accommodating diverse schedules and circumstances.

Assessment and Feedback: Digital tools streamline assessment processes, making it easier for educators to evaluate student performance and provide timely feedback. Online quizzes, automated grading systems, and data analytics help track progress, identify areas of improvement, and inform instructional strategies.

Professional Development: Digitalization provides educators with opportunities for continuous professional development. Online training programs, webinars, and educational platforms offer resources and support to enhance teaching skills, explore new methodologies, and stay updated with the latest research and trends.

Administrative Efficiency: Digitalization optimizes administrative tasks, reducing paperwork, and streamlining processes such as enrollment, scheduling, and record-keeping. Educational institutions can manage student data, monitor progress, and generate reports more efficiently, freeing up time for educators to focus on teaching.

Lifelong Learning: Digitalization promotes lifelong learning by making educational resources accessible to learners of all ages. Online courses, webinars, and Massive Open Online Courses (MOOCs) enable individuals to acquire new knowledge and skills throughout their lives, fostering a culture of continuous learning.

While digitalization brings numerous benefits to education, it is essential to address challenges such as the digital divide, privacy concerns, and the need for digital literacy skills among students and educators. Nonetheless, digitalization continues to revolutionize education, shaping a more inclusive, engaging, and learner-centred landscape.

As an AI language model, I do not have a physical presence or a direct role in the digitalization of education. However, I can provide information on how digitalization is impacting education and the potential roles that various stakeholders play in this process.

Digitalization has had a significant impact on education, transforming the way knowledge is accessed, shared, and acquired. Here are some key roles and aspects related to the digitalization of education:

Educators and Instructors: Educators play a crucial role in leveraging digital tools and technologies to enhance teaching and learning experiences. They may incorporate digital resources, such as online courses, educational apps, multimedia content, and learning management systems (LMS), into their instruction. Digital platforms can enable personalized learning, real-time feedback, and collaborative opportunities among students.

Students: Students are at the centre of the digitalization process, benefiting from increased access to educational resources and opportunities. They can engage in online learning, access educational materials anytime and anywhere, collaborate with peers using digital platforms, and develop digital literacy skills essential for the modern workforce.

Administrators and Institutions: Educational administrators and institutions are responsible for implementing digital technologies, infrastructure, and policies that support digital learning. This involves procuring and managing hardware, software, and online platforms, ensuring data privacy and security, and developing strategies for effective integration of digital tools into the curriculum.

EdTech Companies: Technology companies specializing in educational technology (EdTech) play a vital role in the digitalization of education. They develop and provide learning management systems, online learning platforms, virtual classrooms, adaptive learning software, and other digital tools and resources that facilitate teaching and learning. EdTech companies also contribute to research and innovation in educational technology.

Researchers and Academics: Researchers and academics contribute to the digitalization of education through their studies and insights. They investigate the effectiveness of digital tools, pedagogical approaches, and online learning environments. Their work helps to refine and improve the use of technology in education and inform policy decisions.

Government and Policy Makers: Governments and policy makers establish regulations and policies that shape the digital transformation of education. They allocate resources, promote digital literacy initiatives, address accessibility and equity issues, and develop frameworks for data privacy and security. They also collaborate with educational institutions and stakeholders to ensure a smooth transition to digital learning environments.

9.6 CONCLUSIONS

In conclusion, the integration of AI in personalized learning holds great promise for transforming education. By leveraging AI technologies, educational institutions can create adaptive and tailored learning experiences that cater to the unique needs, preferences, and abilities of individual students. The benefits of AI in personalized learning are manifold. Intelligent tutoring systems offer personalized feedback, explanations, and recommendations, fostering deeper understanding and engagement. Recommendation engines provide students with relevant and personalized learning resources, enhancing their learning experience and motivation. Adaptive assessments adapt to students' progress in real-time, allowing for customized feedback and targeted interventions. However, the implementation of AI in personalized learning also presents challenges. Privacy concerns and ethical considerations must be carefully addressed to protect students' data and ensure responsible use of AI technologies. Additionally, biases within AI algorithms need to be identified and mitigated to ensure equitable learning opportunities for all students. Adequate infrastructure, access to quality data, and teacher training are also crucial for the successful integration of AI in personalized learning environments. To harness the full potential of AI in personalized learning, it is essential for educators, policymakers, and technology developers to collaborate and prioritize the development of robust and ethical AI systems. By doing so, we can create inclusive and student-centric learning environments that empower learners, foster critical thinking skills, and promote lifelong learning.

REFERENCES

[1] Kaswan, K. S., Dhatterwal, J. S., & Kumar, A. (Eds.). (2023). *Swarm intelligence: An approach from natural to artificial.* John Wiley & Sons.

[2] Azevedo, R., & Aleven, V. (2013). *Handbook of metacognition in education.* Routledge.

[3] Baker, R. S., & Inventado, P. S. (2014). Educational data mining and learning analytics. *International Journal of Artificial Intelligence in Education, 24*(4), 387–392.

[4] Conati, C., & Maclaren, H. (2016). Empirically building and evaluating a probabilistic model of user affect. *User Modeling and User-Adapted Interaction, 26*(3), 267–335.

[5] Heffernan, N. T., Heffernan, C. L., & Goldman, S. R. (2014). Using ALEKS to support collaborative learning experiences. In *Artificial intelligence in education* (pp. 413–422). Springer.

[6] Koedinger, K. R., & Corbett, A. T. (2006). Cognitive tutors: Technology bringing learning sciences to the classroom. In *Handbook of educational psychology* (pp. 645–656). Routledge.

[7] Lane, H. C., & vanLehn, K. (2005). Teaching the way, you practice: Combining cognitive and psychomotor skills training in virtual environments. In *Intelligent tutoring systems* (pp. 26–37). Springer.

[8] Martínez-Monés, A., Dimitriadis, Y., & Anguita-Martínez, R. (2017). Learning analytics for the Internet of Things. *Journal of Universal Computer Science, 23*(9), 869–894.

[9] Picard, R. W., & Picard, R. W. (1997). *Affective computing.* MIT Press.

[10] Dhatterwal, J. S., Kaswan, K. S., & Kumar, N. (2023). Telemedicine-based development of m-health informatics using AI. *Deep Learning for Healthcare Decision Making,* 159.

[11] Dhatterwal, J. S., Naruka, M. S., & Kaswan, K. S. (2023, January). Multi-agent system based medical diagnosis using particle swarm optimization in healthcare. In *2023 international conference on Artificial Intelligence and Smart Communication (AISC)* (pp. 889–893). IEEE.

[12] Noor Dayana Abd Halim, Mohamad Bilal Ali, Prof Madya Dr., Noraffandy Yahaya Dr. (2010). Personalized Learning Environment: A New Trend in Online Learning, Conference: Proceeding of Education Postgraduate Research Seminar 2010 (Edupres'10)

[13] Self, J. A., & Roediger III, H. L. (2009). The effects of tests on learning and forgetting. *Memory & Cognition, 37*(4), 501–513.

[14] Kulkarni, C., & Cambria, E. (2020). Artificial Intelligence for personalized education: Challenges and opportunities. *IEEE Intelligent Systems*, 35(1), 64–72.

[15] Woolf, B. P. (2010). *Building intelligent interactive tutors: Student-centered strategies for revolutionizing e-learning*. Morgan Kaufmann.

[16] Siemens, G., Gasevic, D., & Dawson, S. (Eds.). (2015). *Preparing for the digital university: A review of the history and current state of distance, blended, and online learning*. Athabasca University Press.

[17] Wang, M., Shen, R., Novak, D., & Pan, X. (2016). The Impact of Personalization on Student Engagement and Performance. In *Proceedings of the 9th International Conference on Educational Data Mining (EDM 2016)* (pp. 506–511).

[18] Koedinger, K. R., & Aleven, V. (2007). Exploring the assistance dilemma in experiments with cognitive tutors. *Educational Psychology Review*, 19(3), 239–264.

[19] Baker, R. S., Corbett, A. T., Koedinger, K. R., & Wagner, A. Z. (2004). Off-task Behavior in the Cognitive Tutor Classroom: When Students "Game the System". In *Proceedings of the SIGCHI Conference on Human Factors in Computing Systems* (pp. 383–390).

[20] Dillenbourg, P., & Tchounikine, P. (2007). Flexibility in macro-adaptivity: Combining agent and macro-level adaptation. *Journal of Computer Assisted Learning*, 23(1), 1–13.

[21] Johnson, W. L., Rickel, J., & Lester, J. C. (2000). Animated pedagogical agents: Face-to-face interaction in interactive learning environments. *International Journal of Artificial Intelligence in Education*, 11(1), 47–78.

[22] Wang, Y., & Heffernan, N. T. (2019). A review of educational data mining for personalized learning. *Journal of Educational Data Mining*, 11(1), 1–47.

[23] Pardos, Z. A., Heffernan, N. T., Anderson, B., Heffernan, C. L., & Koedinger, K. R. (2006). Using Fine-Grained Skill Models to Fit Student Performance in Educational Games. In *Proceedings of the 8th International Conference on Intelligent Tutoring Systems* (pp. 164–175).

[24] Li, N., Wang, M., Chen, W., & Gao, J. (2019). Personalized learning for K-12 students: A review of the literature. *Computers & Education*, 137, 209–225.

Chapter 10

Transformative innovation in education

Sapan Adhikari
Managing Director, SST.Pvt.Ltd, Attariya, Kailali, Nepal

10.1 OVERVIEW

Education is a learning process where one is involved in acquiring or teaching wisdom for self-development in order to rationalize, judge, and make decisions towards the things happening around one's life. It helps to prepare someone to be intellectually mature in his/her life [1]. The word education came from the Latin "Educare", meaning "to nourish" or "to raise," and Educatus, which translates as education [2]. When one undergoes education, it often happens in formal or informal settings. Formal education is a kind of official learning process which is usually divided into stages like preschool or kindergarten, primary school, secondary school, and then college, university, or apprenticeship while informal education happens outside of a structured curriculum like unschooling or homeschooling, autodidacticism (self-teaching), and youth work. The science or art of teaching or the methodology is called pedagogy.

Traditionally and up till now, education happens in a closed setting or institutions like schools or colleges. A teacher teaches the students with their predefined curriculum or syllabus. The teaching method includes reading books, explanation, writing, visualizing, thinking, and written tests to assess if the students are learning properly or not.

Along with the Industrial Revolution 1.0 of the steam engine invention to the world of AI, IoT and green sustainable development of 5.0, the human civilization has evolved immensely. This has happened because of the same pace of development happened in education i.e., 1.0 to 5.0. This has resulted in a continuously growing trend in technology, along with unceasing demand for sharp skilled and problem-solving manpower causing an ever-increasing demand for innovation in education.

10.2 EVOLUTION OF EDUCATION

Education has evolved over the centuries from teaching beliefs, and religions to the modern-day classroom, schooling, pedagogy, and digital devices and can be seen in two forms i.e., traditional education and modern education.

10.2.1 Traditional education

Traditional education or customary education or conventional education is all about imparting knowledge about the values, manners skills, and social practices to the student or apprentice. In traditional education oral recitations were given to the student about the customs and traditions of the society in which he/she lived [3].

In traditional education, the process would occur as the students simply sitting down together and listening to the teacher or another who will recite the lesson. Assessment was usually done through oral tests, not as written tests as we do in current schooling. Science and technology would not be discussed and only customs, traditions, and religions were key aspects of discussion, which is why it is called traditional education [3].

10.2.2 Modern education

Modern education is very different from traditional education which is comprised of learning plans or pedagogy, assessment, science, technology, skills development, etc. Modern-day schools, colleges, universities, and online training centres are the best examples of modern education. Modern education focuses on listening, writing, visualizing, imagining, thinking skills, and written assessment to check if the students are learning properly or not. Modern education is just an evolution of traditional education [3]. Evolution in modern education can be segregated into further stages as follows.

10.2.3 Education 1.0

Most of the current-day schools and colleges function as the Education 1.0 model. They have a fixed set essentialist-based curriculum which guides the way of teaching and testing to the teacher. The curriculum is often based on traditional disciplines such as mathematics, science, social, history, language, and literature. The teacher is the controller of the class and decides what is important to teach regardless of the student's interests. The teacher is the main body to assess and evaluate the student's performance. The classroom setting looks like students sitting in rows at desks and learning in mass and listening to the teacher [4].

Education 1.0 is a largely one-way process in which students go to school to get an education from teachers and teachers teach them on a routine basis. The teacher provides class notes, handouts, textbooks, and videos to students for further improvement of assessment and their development. The students take the information gathered from the teacher and get prepared for the assessment. This is much centred towards teachers [4]. (Figure 10.1)

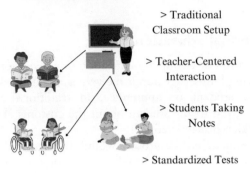

Figure 10.1 Education 1.0 in a nutshell [4].

10.2.4 Education 2.0

Education 2.0 is more interactive than Education 1.0 where the teachers and students are more engaged and communicate on less understood topics and students among themselves interact with content or experts to get a better understanding of what they are learning. The schools or colleges provide extra activities or industry exposure like industry visits, project-based learning, corporate expert guest lectures, etc. to the students so the student can get extra learning from the curriculum. The school encourages blog postings, Wiki reading, and social networking in the classroom through Facebook, WhatsApp, and Skype for better sharing of information among the students. Overall, it is student-centric learning, learning-based outcomes, and result oriented. But still, most schools and colleges fail to practice Education 2.0 as it resides on paper only [4]. (Figure 10.2)

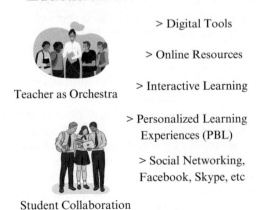

Figure 10.2 Education 2.0 in a nutshell [4].

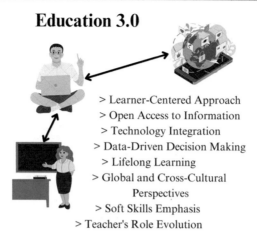

Figure 10.3 Education 3.0 in a nutshell [4].

10.2.5 Education 3.0

Education 3.0 is based on problem-solving, creativity, and innovation. Students are provided with free and readily available content so they can strive for self-direction and interest-based learning. The ultimate purpose of Education 3.0 is to produce industry-ready graduates or students [4] (Figure 10.3)

10.2.6 Education 4.0

Using Education 3.0 as its basis, Education 4.0 is introduced and it has more emphasis on collaborative and personalized learning. It enables students to choose a mode of higher education for their career path as per their interest and the student can learn on campus, at home, and even in the workplace. Education 4.0 is often termed as "Smart" as the internet and modern smart digital devices like smartphones, tablets, laptops, computer applications, and technology are used widely for tutorials, assignments, and assessments. Higher Education 4.0 goes beyond the boundaries of higher education institutions and the learner of 4.0 has full freedom to choose courses, knowledge, and skills from different platforms like multiple universities offering choice-based courses and acquiring degrees of their choice. This is pure outcome-based education [5].

The key factor in Education 4.0 is the MOOCs from different governmental and non-governmental universities, institutions, etc. (Figure 10.4)

10.2.7 MOOCs

Massive Open Online Courses (MOOCs) are free online courses from a wide range of career streams available for anyone to enrol in different

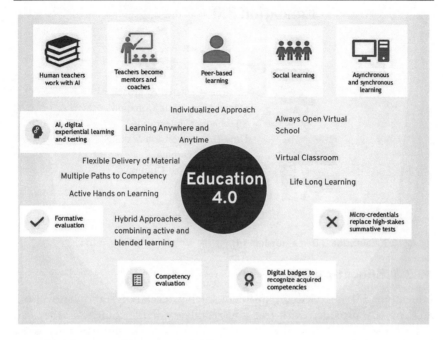

Figure 10.4 Education 4.0 in a nutshell [6].

reputed universities or private institutions like MIT, Harvard, etc. MOOCs provide an affordable and flexible way to learn new skills. It helps one to improve and advance their career in their desired field of interest by delivering quality educational experiences at scale. Millions of people around the world use MOOCs to acquire knowledge for career development, changing careers, college preparations, supplemental learning, lifelong learning, corporate e-learning and training, and more [7].

MOOCs have dramatically changed the learning pattern of students. The traditional was limited to time and space but the MOOCs are flexible and way beyond that. MOOCs or massive open online courses open a new opportunity for students to grab education for free [7]. There are usually no barriers or entry requirements for enroling in MOOCs. You can enrol and get an education regardless of the place you live or your financial circumstances [8]. (Figures 10.5 and 10.6)

Some of the popular MOOC platforms are given in Figure 10.6.

10.2.7.1 How does a MOOC work?

MOOCs usually start with enrolment by giving personal information and the course will have a specific start and finish date. As the MOOC starts, you'll be provided with a wide range of interactive tools to interact with university educators and other learners. These tools include video lectures,

M	Massive	> Millions of User > Mass Community
O	Open	> Open Contents > Open Registrations > Free of Charge
O	Online	> Real-time Interactions > Online Learning
C	Course	> Diverse Courses > Start and End Date > Self Paced Learning > Instructor's Role > Scripted Assessments and Feedbacks > Certification

Figure 10.5 MOOC in a nutshell [9].

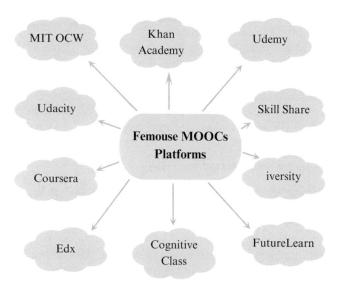

Figure 10.6 Popular MOOC platforms [10].

articles, discussions, assignments, and social networking. For getting help there will be community or blog posting where discussions will be conducted on the problems or topics. Millions of learners and educators take part simultaneously. As you progress through the course there will be some pre-defined assessments given and the assessment will be via peer-reviewed written assignments or computer-marked tests, rather than by

124 Transformative innovation in education

Figure 10.7 Free machine learning course offered by Stanford University on Coursera platform [12].

tutors. In short, MOOCs are free courses which take place completely online [11].

In Figure 10.7, the most popular free machine learning course among the data science aspirants is presented which is available on the Coursera platform. Any enthusiast having no educational background can enrol freely and can study machine learning. This course is offered by Stanford University and after completion the university provides a certification of completion.

This is one of the examples of MOOCs and there are many more on other platforms like MIT open course, Udemy, etc.

10.2.8 AI in personalized learning

In traditional schools, lessons, tutorials, and assignments are already set for a year where teachers are guided with what they are supposed to teach and what students are supposed to be learnt in that particular year. Whether the students like it or not or are interested or not to study, they must go through that fixed set of pedagogical instruction in order to pass that academic year. This kind of approach is known as "one size fits all" where every student is taught in the same way despite every student has their own unique interest and skill sets. And because of this, many students are seen to fall aside in their career. As a counter, a new learning process is emerging and becoming popular which is known as Personalized Learning [13].

Personalized Learning is a process in which a student is taught a stream or field in a way which they are keen on and aspire to learn. Someone might want to be a doctor or computer scientist or sportsperson or musician so to cater to this every student is given a learning plan based on their areas of interest, skills, and learning pace. This helps the student to take ownership of their learning process, build self-advocacy, and encourages them to engage their interest [13]. Examples of personalized learning are separate music schools, corporate training, online courses like udemy.com, skillshare, MIT open learning library, etc. [13].

10.2.8.1 How personalized learning works?

Personalized learning sets a rigorous standard and expects high-quality learning outcomes from the students. There are basically four types of methods that schools follow.

1. Keeping track and record of the student's profile that provides full-fledged information about the student's interests, strengths, needs, assessment, progress, and goals.
2. Based on that track record of students the school can use it to customize the learning plan and schedule for the students that help them to progress, motivate, and achieve their goals. Several learning methods or modalities can be used such as project-based learning, independent project work, complex task assignment, and one-to-one tutoring with teachers.
3. Schools can use a competency-based progression like an assessment review from different students and monitor the progress of their specific goals. This helps to segregate and identify the students from their ability to learn at what pace and who is lagging behind in the process. This also helps to identify the core strength of the students and work towards mastery in their field of interest.
4. Schools or institutes can adopt flexible learning environments like the physical setup of classes, allocations of teachers, time schedule allocations so that the students can best learn [13].

10.2.8.2 Artificial intelligence and Personalized learning

Artificial intelligence and personalized learning are a combination of data, machine learning, and analytics in an ed-tech product. Applications (web, desktop, or app-based) are trained with a large set of data to perform some teaching tasks on a large scale. When a student asks or search for information it provides the results in an instant with much higher accuracy. For example, ChatGPT from Microsoft and Bard from Google where any task given related to information or asking a question or even drawing a picture from text direction gives results instantly. Students can ask for essay writing, engineering topic question, mathematical question, geographical question, or even geo-political question and it can provide accurate information or answer in no time.

In the MOOC program or in eBooks from eLibrary, AI can be seen in action. The interface shows in which chapter the students are in with accurate time, when they left their video tutorials, what they have studied from the resources provided, and what is left to study. Along with that assessment reviews and proper feedback-providing systems which highlight the key areas to focus on in their study are shown in the interface in the system. AI programs also choose the right digital content at the right time. The AI-embedded system helps the student's learning experiences become more personal and successful.

For example, A Altitude learning platform helps the teacher to create individual learning paths for each student based on the student's learning pace and monitors students' progress at the same time. The software DreamBox Learning helps to increase math knowledge by specifically providing problems according to the individual student's strengths and weaknesses and slowly helps the student to build up expertise. AI can be used not only to aid students but also to provide guidelines for the educator's or teacher's job, e.g., an automated dashboard interface helps to create scoring, review, and feedback for student performance [14].

10.3 EMERGING TRENDS AND TECHNOLOGY IN HIGHER EDUCATION

1. Competency-Based Learning/Education, where technology is used for personalized learning and is student-centred revolving in their field of interest, skills, and passion.
2. Video Streaming/Flipped Classroom/eLearning trends from Zoom to Skype to Webinars and even live streaming on social media itself. Video was perhaps the most visible and common form of technological and educational trend during the COVID-19 pandemic. The flipped classroom seems to be in our pockets, as so much great content is already published and accessible on YouTube and other platforms.
3. Open curriculum, like MIT's OpenCourseWare and other MOOCs.
4. Digital textbooks or elibrary.
5. Use of data analytics to check the student's interests, performances, results, etc.
6. Virtual, Augmented Reality, and Contextual Tools to visualize learning objects.
7. Extended reality (XR) helps the students move away from traditional lectures towards more engaging, immersive learning experiences within a simulated real-world space. Other benefits include increased comprehension levels and long-term memory retention among students.
8. Gamification learning through the game model using different software platforms. This enables greater student interaction and more excitement about learning.

9. New kinds of certifications and degrees are helping "update" certifications and degrees over time.
10. New pedagogies, like project-based learning at the college level, inquiry-based learning, competency-based learning, experiment-based learning, and more [15,16].

10.4 EDUCATION 5.0

In traditional educational approaches, strategies, and pedagogy, each year the quality of graduates produced continues to deteriorate in terms of skills and knowledge. This causes a huge gap between industry demand and graduates' knowledge and skills [17].

So, Education 5.0 comes into the picture. It is a transitional approach from theoretical-based education to action/outcome-based systems. Education 5.0 is not just about teaching, research, and being employed but to innovate and industrialize. It is all about problem-solving for value-creation at his/her/learner's stage of education so that later on they will not stumble when choosing what is the right thing to do and what is not for their career. (Jonathan, 2019)

Education 5.0 is possible because of the adoption of technology like the Internet of Things (IoT), advanced robotics, automation, etc. The main emphasis is on personalized learning, information technology, easy access to resources, flexibility, continuous assessment and improvement, critical thinking, being creative, problem-solving, student participation, and project based learning [18].

Education 5.0 focuses on humans and uses technology for creating efficient and effective graduates. It is more about being rational and making conscious decisions while seeing the bigger picture. it takes a holistic approach or takes consideration of all stakeholders such as educational institution, government, industry, community and learner themselves to provide a holistic approach to provide what is needed in education for the students or learner. The ultimate objective is to develop an individual intellectually, socially, and emotionally strong by applying strategic methodological and pedagogical approaches. The motive here is to motivate, make creative, innovative, and joyful learning experiences for the learners so they can have self-esteem and confidence for lifelong learning [19].

10.5 FUTURE OF EDUCATION

10.5.1 E-learning

When the COVID-19 pandemic hit, all of the educational institutions around the world shut their classes, affecting over 1.2 billion children. As a result, education was transformed into the e-learning platform where teachers were conducting classes on the digital platform remotely and students were taking classes on the same platform and doing assignments at the same times. With this new trend, many ed tech-based IT companies were established and are still emerging [20].

With this, a shift from the traditional classroom to the digital platform happened around the world. Many education experts doubt if this trend will remain post-pandemic; however, this education paradigm shift has impacted the worldwide education market and helped the smooth running of education institutions [20].

If we care to look, before COVID-19 and the transformation of the education system, high growth in the adoption of new education technology was already clearly emerging. The global edtech investment in 2019 was US $18.66 billion and this is estimated to be a market of $350 billion by 2025 in online education projects. During COVID-19 the usage of language apps, virtual tutoring, video conferencing tools, or online learning software significantly surged [20].

While some believe this unplanned rapid transition into the digital platform with no training, insufficient bandwidth, and little preparation will result in a poor user experience and would not be sustained in the future, others believe a new hybrid model with traditional and new education systems will have significant benefits. The setback of e-learning is not seen yet but there are already success stories surfacing regarding the

Figure 10.8 What work conference meetings currently look like in Facebook's metaverse [22].

transitions amongst many universities. For example, Zhejiang University was able to develop more than 5,000 online courses in two weeks' time for the transition using "DingTalk ZJU". The Imperial College London commenced offering an online course related to the science of coronavirus, which has been the most enroled class launched in 2020 on Coursera [20].

Dr Amjad, a Professor at the University of Jordan who has been using Lark to teach his students said, "It has changed the way of teaching. It enables me to reach out to my students more efficiently and effectively through chat groups, video meetings, voting, and also document sharing, especially during this pandemic. My students also find it is easier to communicate on Lark. I will stick to Lark even after coronavirus. I believe traditional offline learning and e-learning can go hand in hand" [20].

Figure 10.9 Learning in the metaverse 1 [22].

10.5.2 Metaverse

On October 28, 2021, Facebook CEO Mark Zuckerberg announced Meta, a brand name for his company which would encompass their core apps into one company. The renaming into Meta was a transitional commitment to developing the metaverse of a virtual social world where everyone is connected and interacts just as we are connected in real life. The idea is to create a more immersive internet world where technology like AR and VR will be used to spend our time engaging in virtual spaces and experiences rather than the physical world. The term was first coined by Neal Stephenson in 1992 in his science fiction novel Snow Crash [21].

In this race to develop a metaverse world many companies have come forward to develop their own metaverse world like Microsoft, Facebook, Nvidia, Apple, Niantic, Decentraland, etc. With the rollout of 5G or 6G,

Figure 10.10 Learning in the metaverse 2 [22].

10G cable, fast internet through Starlink satellites, and other low latency technologies, it has been a hot topic to discuss among these companies [21].

In the form of development in education, it can be seen as a virtual setup like a classroom or meeting hall, a school where no matter where each student is, they can be connected to this virtual environment setup with their own 3D avatar, VR tools, and headphone to attend the class or meeting or assembly just like they do in the physical school in the real world. They could see all the participants hanging out in the same room, and interact with each other seamlessly which is not possible in video calls of any app like Skype, Messenger, Zoom, etc. (Figure 10.8).

If one who wants to learn anything, he/she could bring all the subjects of interest closer to him/her. For example, someone who wants to learn astrophysics and learn about the solar system, he/she can bring a real virtual

Figure 10.11 Teleporting to ancient Rome I [22].

setup upfront to him/her and can study wisely in all the dimensions he/she wants. (Figure 10.9 and 10.10)

Or suppose someone wants to learn about the history of the Roman empire, he/she could teleport to not just the place but time and could witness moments from 2,000 years ago. (Figures 10.11 and 10.12)

With the right headset or glasses, one can pull out or pull up schematics of their study. (Figure 10.13)

For a medical student or a doctor, he/she can learn a new technique of surgery by firsthand practice until they get it right. (Figures 10.14 and 10.15)

10.5.3 AI-based Chatbots (ChatGPT and Bard)

Artificial intelligence-based chatbots are chatbots to create and chat like a human conversing with another human. This chatbot responds to every

Figure 10.12 Teleporting to ancient Rome 2 [22].

Figure 10.13 Pulling out schematic of study in metaverse [22].

question asked, similar to a human responding from the back. It is built by using multiple combinations of machine learning models like natural language processing, neural networks, supervised learning, and reinforcement. In order to operate it used millions of millions of datasets and trained itself to respond to any questions asked. Using these chatbots people can ask for writing code for computer programs, compose music using a certain keyword, solve math problems, describe complex topics in a simpler form, search for a job and write resumes in a simple form, etc. The most popular AI-based chatbots are ChatGPT by Microsoft and Bard by Google [23]. (Figures 10.16 and 10.17)

134 Transformative innovation in education

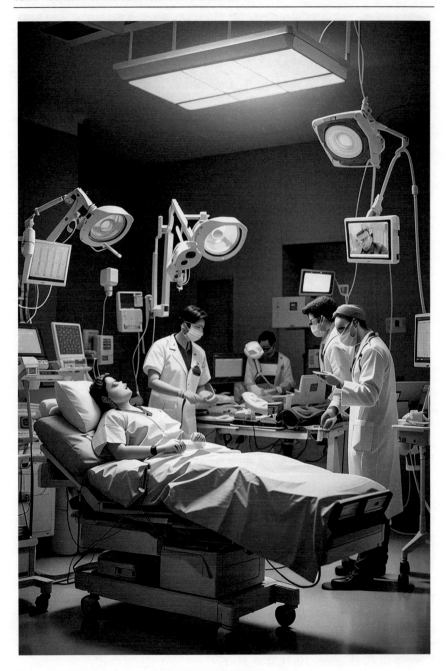

Figure 10.14 Learning surgery in metaverse 1 [22].

Advances in technological innovations in higher education 135

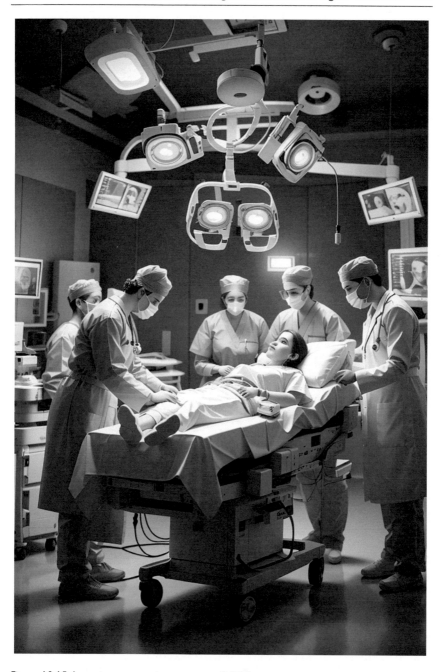

Figure 10.15 Learning surgery in metaverse 2 [22].

Figure 10.16 Sample interface of asking ChatGPT to write a song on a mountain [24].

Figure 10.17 Sample interface of asking Google Bard about algebra [25].

10.6 CONCLUSION

As industry is shifting from Industry 4.0 to 5.0 of IoT, robotics, automation, etc., it is now becoming imperative for each country in the world to shift their educational format to Education 5.0. Otherwise one's country's economy could never flourish as it will never be able to produce effective and efficient skilled manpower. Skilled manpower is key to innovation, startup, industry growth, etc. The main focus for them should be on the theme of "Innovating Education and Educating for Innovation".

The citizens' quality of education defines a country's social and economic well-being. A higher "knowledge society" means a more skilled labour force and this makes citizens of the country more efficient and effective.

This helps to achieve the common goal of the nation in a holistic way like an increase in life expectancy, per capita income, increase in living standard, etc. As the education paradigm shifts to a progressive chain, there is much need for innovation, the inclusion of stakeholders, the use of digital technology, the full utilization of resources, and more focus on outcome-based learning.

When we try to innovate education, there is one thing that never should be left behind i.e., the ethics of the students. The learning should always encompass positive social conduct, values, and ethics and these should always be instilled in the student's hearts. The ultimate goal of innovative education and education innovation is to make students or learners possess "career readiness" to "life readiness". Teachers should also be trained in parallel to cope with the new Education 5.0 so that they could deliver more effective and efficient knowledge to the students.

REFERENCES

[1] Dictionary. (2012). www.dictionary.com. Retrieved from Education: https://www.dictionary.com/browse/education
[2] Encyclopedia, N. W. (2019). *Education*. Retrieved 12 21, 2021, from https://www.newworldencyclopedia.org/entry/Education
[3] Anchal, M. A. (2018). *Traditional Education System versus Modern Education System: A reference to Indian Education system*. Retrieved 12 21, 2021, from https://madhavuniversity.edu.in/reference-to-indian-education-system.html
[4] Gerstein, J. (2018). *The Difference Between Education 1.0 & 3.0*. Retrieved from www.teachthought.com: https://www.teachthought.com/the-future-of-learning/past-time-education-3-0/
[5] Gavhane, D. S. (2019). *Higher Education: A Journey from 1.0 to 4.0 (PART 2)*. Retrieved from Eduvoice. in: https://eduvoice.in/%EF%BB%BFhigher-education-a-journey-from-1-0-to-4-0-part-2/
[6] Leger, M.-A. (2019). *An introduction to Education 4.0*. Retrieved 12 21, 2021, from https://www.slideshare.net/maleger/an-introduction-to-education-40
[7] MOOCs. (2021). *About MOOCs*. Retrieved 12 21, 2021, from https://www.mooc.org/
[8] e-learningmatters. (2022). *CMOOCS VS XMOOCS: The Battle of the Pedagogies*. Retrieved 12 21, 2021, from https://www.e-learningmatters.com/cmoocs-vs-xmoocs-battle-pedagogies/
[9] Nisha, F. (2018). *Massive Open Online Course*. Retrieved 12 21, 2021, from https://www.researchgate.net/figure/Massive-open-online-course-MOOCs-3_fig1_273888190
[10] onlinecoursereport. (2021). *The Best Tools for Developing Online Courses*. Retrieved 12 21, 2021, from https://www.onlinecoursereport.com/the-best-tools-for-developing-online-courses/
[11] FutureLearn. (2016). *What is a MOOC?* Retrieved 12 21, 2021, from https://www.futurelearn.com/info/blog/what-is-a-mooc-futurelearn

[12] coursera. (2021). *Machine Learning*. Retrieved 12 21, 2021, from https://www.coursera.org/learn/machine-learning#instructors
[13] Morin, A. (2021). *Personalized Learning: What You Need To Know*. Retrieved 12 21, 2021, from https://www.understood.org/articles/en/personalized-learning-what-you-need-to-know
[14] Focke, C. (n.d). *Talking Teaching: Is Personalized Learning the Future?* Retrieved 12 21, 2021, from https://www.adinstruments.com/blog/talking-teaching-personalized-learning-future
[15] Heick, T. (2021). *14 Examples Of Innovation In Higher Education*. Retrieved 12 21, 2021, from https://www.teachthought.com/the-future-of-learning/innovation-higher-ed/
[16] Council, F. T. (2019). *10 Ways Edtech Advances Are Shaking Up Education*. Retrieved 12 21, 2021, from https://www.forbes.com/sites/forbestechcouncil/2019/12/10/10-ways-edtech-advances-are-shaking-up-education/?sh=634c64c53cb2
[17] Meenakumari, D. (2021). *Response to Demands From Society Through Education 5.0 in Indian Education System*. Retrieved 12 21, 2021, from https://timesofindia.indiatimes.com/readersblog/response-to-society/response-to-demands-from-society-through-education-5-0-in-indian-education-system-2-33163/
[18] Uysal, D. L. (2021). *Education 5.0*. Retrieved 12 21, 2021, from https://www.leventuysal.com/2021/03/28/education-5-0/
[19] Dervojeda, K. (2021). *Education 5.0: Rehumanising Education in the Age of Machines*. Retrieved 12 21, 2021, from https://www.linkedin.com/pulse/education-50-rehumanising-age-machines-kristina-dervojeda/
[20] weforum. (2021). *The COVID-19 Pandemic has Changed Education Forever. This Is How*. Retrieved 12 28, 2021, from https://www.weforum.org/agenda/2020/04/coronavirus-education-global-covid19-online-digital-learning/
[21] Schroeder, R. (2021). *Tech Trends in Higher Ed: Metaverse, NFT and DAO*. Retrieved 12 12, 2021, from https://www.insidehighered.com/digital-learning/blogs/online-trending-now/tech-trends-higher-ed-metaverse-nft-and-dao
[22] Meta. (2021). *The Metaverse and How We'll Build It Together - Connect 2021*. Retrieved 12 21, 2021, from https://www.youtube.com/watch?v=Uvufun6xer8
[23] Hetler, A. (2023). *Bard vs. ChatGPT: What's the Difference?* Retrieved from www.techtarget.com: https://www.techtarget.com/whatis/feature/Bard-vs-ChatGPT-Whats-the-difference
[24] OpenAI. (2023). *ChatGPT*. Retrieved from https://chat.openai.com/: https://chat.openai.com/
[25] Bard. (2023). *Bard*. Retrieved from https://bard.google.com/: https://bard.google.com/

Chapter 11

eSCOOL

A virtual learning platform

Ruqaiya Khanam[1,2], Shraiyash Pandey[3], Shrishti Choudhary[3], and Abhik Kumar De[3]

[1]Department of Electronics and Communication Engineering
[2]Center for Artificial Intelligence in Medicine, Imaging & Forensic, Sharda University, Greater Noida, India
[3]Department of Computer Science and Engineering, Sharda University, Greater Noida, India

11.1 INTRODUCTION

While there has been much research on data compression, video quality handling, and connectivity issues in video conferencing apps, not many researchers have taken into account or addressed the regularly faced problems inside the meeting for online school classes. It is not the case that data compression or video quality handling topics are less relevant or not as important but in some areas, these additional concerns are very important in an online meeting experience which is currently not being discussed. In fact, web conferencing systems are, if not entirely, then mainly dependent on the internet and the connectivity for online communication. But specifically referring to the teacher-student classroom environment, there are several problems that arise in a daily online classroom setting. Many problems, drawbacks, and issues that are currently being faced in already existing video conferencing apps are undiscussed and disregarded. These problems need to be addressed and alternative solutions have to be proposed to make the online video conferencing experience much better. Specific problems faced during these online school classes are related to attendees list, less centralized and restricted access in meetings for teachers, and unavailability of certain features to improve and enhance the teaching-learning experience. Technology is very initiative and helpful that can resolve such problems. Creating a video conferencing app that not only provides already existing features in other video conferencing apps but also focuses primarily on including new features that provide alternative solutions to these pre-existing problems is the main goal. Such an app will make both teachers' and students' educational teaching-learning experience more reliable and fun. [1–6]

DOI: 10.1201/9781003376699-11

The body of the survey provides a brief and detailed comparative analysis of the differential gap between various pre-existing video conferencing apps. It highlights the areas that are left undiscussed and unresolved. Highlighting these main areas helps provide an overview of the necessary elements in the new video conferencing app. afterwards, an analysis of parameters and features within video conferencing apps is discussed. These parameters provide a pleasant and reliable experience for online video conferencing and meetings. Each parameter is heavily discussed and analyzed so the audience can have a better understanding of this video conferencing app. These features are also highlighted to differentiate the gap between different video conferencing apps which will be addressed in this video conferencing app.

11.2 THE ENVIRONMENT OF ONLINE CLASSROOM

It is no mystery that an online classroom is not the same as an offline classroom. In fact, during the COVID-19 pandemic, the whole point of the online classroom was to replace the offline classroom until the situation went back to normal. It was never meant to be a permanent replacement since it does not provide the same experience as an offline classroom would. However, after implementing the online classroom into place for a while, people have realized it also comes along with many advantages compared to offline classrooms. An attendance system that requires almost no effort to use for teachers can be highly beneficial and effective. However, in order for this system to be effective, it can only be implemented in an online classroom. Similarly, a feature to organize notes efficiently and in a much easier way can be done in an online teaching-learning environment. Only a highlight of features can be stated, but many more can be implemented in the future with the advancement of this technology. Many think of an online classroom to be inefficient or useless, but many students tend to be more productive learning via online sources. Not to mention, many successful people are in a very great position with the help of online resources. The environment of an online classroom can be as effective, and beneficial if not more than an online classroom. (Figures 11.1 and 11.2)

11.3 IMPORTANCE OF ESCOOL

A student's productivity is shortened when the right environment is not provided. A place that provides students with proper and effective education is vital and necessary. eScool makes a student feel comfortable and flexible with online learning. It is not only helpful for the students but

Advances in technological innovations in higher education 141

Figure 11.1 Environment of an online classroom.

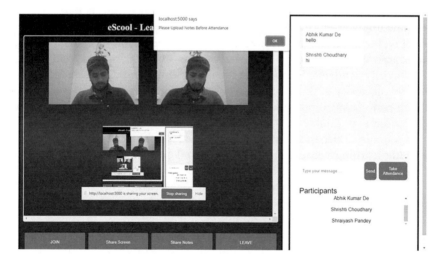

Figure 11.2 Notes uploading facility before attendance.

also for teachers. It provides teachers with handy features that reduce time and effort, therefore, they can focus more on the educational aspect. To enhance and improve both the student and teacher's classroom experience, eScool is an important tool that solves a major issue of productivity in online learning for students. It becomes difficult for students to focus and concentrate in class, especially when there are so many unwanted distractions. To improve the quality of the online teaching-learning experience, a platform with effective features can remove these unwanted distractions and increase students' ability to concentrate and focus.

Our work outlines a vision to provide students with the best online video conferencing experience. Many different pre-existing video conferencing applications have features that are very developed and modified to the best extent. However, there are some areas that can be improved or developed. Accordingly, we extracted limitations, problems, and issues being faced in pre-existing video conferencing apps to develop and improve upon each area or field that these apps lack. Overall, our online video conferencing app, eScool, provides certain features that address the limitations of currently used, worldwide known video conferencing apps such as Zoom, Google Meet, and Microsoft Teams [7].

11.4 METHODOLOGY

A video conferencing app is a software tool that lets you conduct video meetings online via video conferencing, audio conferencing, and screen sharing.

Many applications like Google Meet and Skype already existed and were being widely used for remote conversations, especially in business meetings and corporate discussions. However, it was only with the advent of the coronavirus pandemic when it became impossible to hold physical meetings that the necessity and use of video conferencing apps grew exponentially.

The major contributor to this rise in the use of video conferencing apps like Zoom is undoubtedly "Online Education" [8–11].

As schools, colleges, and universities were forced to shut down, the entire teaching-learning process had to be adjusted and transformed into online mode. The use of multimedia and real-time content sharing compensated for the loss of physical interaction in offline mode, but there were still nonetheless many bottlenecks and loopholes in most of the existing video conferencing applications like Zoom.

Every time there is a slight hint the pandemic is over, a new variant is found. People across the world cannot achieve a bright future because of the lack of proper education. Our invention provides an easily accessible virtual platform for teachers and students to interact efficiently across the globe. Although there are already existing apps that target this worldwide problem of lack of education, we focus primarily on the drawbacks and limitations of these existing apps. Such platforms lack in some areas to provide or deliver an efficient teaching-learning environment. We present a platform that tackles these issues. We hope to see a revolution in this field of the online education system as time progresses such that future generations can be provided with an excellent form of education no matter where the school is located. Therefore, the platform is called eScool (Figures 11.3 and 11.4).

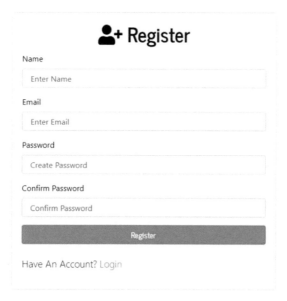

Figure 11.3 Registration window.

11.5 PROBLEM – SOLUTION REPRESENTATION

Drawbacks (of existing apps) and solutions:

- Restricted access
  - **Problem:** Ease of access allows miscreants to enter the classroom with organizational IDs by mimicking actual students' names and cause disturbance
  - **Solution:** Only the students pre-enrolled with their accorded IDs by the faculty itself can access the classroom via link

144 eSCOOL: a virtual learning platform

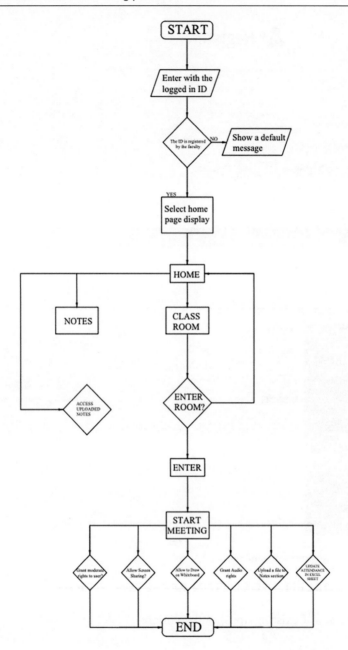

Figure 11.4 Flowchart of walkthrough in app.

Advances in technological innovations in higher education   145

- One-click attendance
  - **Problem:** So much chaos and wastage of study time are caused while taking daily attendance
  - **Solution:** eScool allows teachers to record attendance in a matter of one click

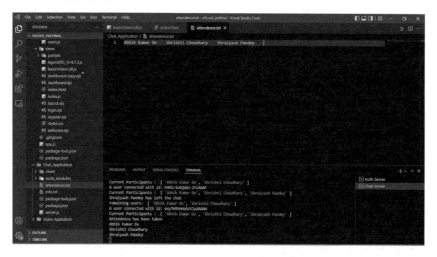

- Easy notes sharing
  - **Problem:** Numerous times the faculty is unable to upload or share study materials which hampers the teaching-learning process.
  - **Solution:** While presenting the study materials in an online class, eScool provides a feature that enables the faculty to upload and share the same file instantly with just one click

- Safety Pocket Mode
  - **Problem:** Numerous times participants face embarrassment or cause unintended disturbance due to accidentally switching on or off their camera or microphone
  - **Solution:** eScool provides a handy screen lock, Safety Pocket Mode, which prevents such embarrassment
- Single link classroom
  - **Problem:** Too many links shared to students makes it confusing and difficult for them to keep up
  - **Solution:** Our platform simulates real-life classrooms in a virtual mode wherein there is a single link for each class and respective teachers join at the time of their class
- Student's Personal notes
  - **Problem:** Students have to switch between third party apps such as Google Docs while taking quick digital notes during lectures
  - **Solution:** eScool provides one step integrated solution to this issue by introducing a feature that allows students to take quick digital notes within the app during ongoing lecture
- Scheduled Meeting reminder
  - **Problem:** Even after attaining a timetable of classes it becomes difficult to keep track of upcoming lectures and their timings on a daily basis
  - **Solution:** With eScool, there are no more worries regarding an important lecture. A reminder is sent at an appropriate time before the class is scheduled

11.6 SIMILAR ONLINE APPLICATIONS

Comparative analysis of existing video conferencing applications:

There is a lot of news related to which video conferencing app is best to use today. The apps available on the market are witnessing a huge rise in popularity and use, especially due to lockdowns across the globe. These apps do not target a specific audience, in fact, the market space stretches from as small as a kid in 2nd grade to working professionals. It can be used for online calls or just casual group video calls. Even though there is a lot of competition in this market space, each app has similar features with few exceptions. The three most popular apps used widely across the world are Zoom, Google Meet, and Microsoft Teams [7].

11.6.1 Zoom

Zoom is currently known to be the most popular video conferencing app. The app has over 200 million active users on a daily basis. Zoom has amazing features, along with a few drawbacks. It has extremely good features like screen sharing, screen recording, and team chats. One of the most amazing features is participants can change their background with default options or put up an image of their choice. (Table 11.1)

11.6.2 Google Meet

Google Meet, previously known as Hangouts, is also one of the trending video conferencing apps on the market. It is free of use, which provides a smooth interface and elegant look. The main target audience for this app is companies and educational centers. Google Meet is connected to Google Suite, which allows participants to add meetings through Event Calendar. Google Meet is popular due to the recognition of the Google name itself. In fact, it doesn't have as many features compared to the other two apps, Zoom and Microsoft Teams. (Table 11.2)

Table 11.1 Major Zoom tools

Features	Usage
Schedule a meeting	To schedule a meeting at a specific time
Use of the calendar	To receive notifications of meetings that are scheduled
Screen sharing	Participants have the option to share any sort of information via this feature with other participants in the meeting
Screen Recorder	Allows participants to review anything after the meeting is over especially for students who need to revise a topic after it has been taught
Virtual whiteboard	Allows participants to express ideas on a whiteboard by drawing, writing, or carrying out explanations
Chat	Participants have the option to interact both directly and privately
User management	You have the option to enable and disable the audio and video of the participants

Table 11.2 Major Meet tools

Feature	Usage
Schedule a meeting	To schedule a meeting at a specific time
Use of the calendar	To receive notifications of meetings that are scheduled
Screen sharing	Participants have the option to share any sort of information via this feature to other participants in the meeting
Screen Recorder	Allows participants to review anything after the meeting is over especially for students who need to revise a topic after it has been taught
Virtual whiteboard	Allows participants to express ideas on a whiteboard by drawing, writing, or carrying out explanations
Chat	Participants have the option to interact both directly and privately
User management	Options available to enable and disable audio and video of participants as host

11.6.3 Microsoft Teams

Microsoft Teams is part of Microsoft Office 365. There are two different types of plans available: free and paid. This virtual meeting platform is also used by hundreds of students nationwide. The app focuses mainly on collaboration and teamwork. Collaboration and teamwork are key in an online classroom to achieve the best workflow. The app is not only just an online video conferencing app, it also includes features that are not available in other apps. Microsoft Teams is an all-encompassing, robust tool for collaboration and sharing that allows staff from multiple locations to work together as a seamless, functional unit [7]. Stability, security, privacy, and easy access are some of the hallmarks of Microsoft Teams. (Tables 11.3 and 11.4)

Table 11.3 Major MS Teams tools

Tools	Usage
Screen sharing	Allows all participants to have the option to choose what to share with other meeting participants
Integrated Classroom	Students can easily access recordings and notes from the session within the same platform
Whiteboard	Allows participants to express ideas on a whiteboard by drawing, writing, or carrying out explanations.
Virtual Backgrounds	Allows users to bring variety and creativity into play by expressing different types of backgrounds

Table 11.4 Overall comparison of existing competitor works cited

Features	Zoom	Google Meet	Microsoft Teams	eScool
Restrictive access to pre-enrolled students with respective email-id.	✓	✗	✗	✓
A feature to download or extract the attendance of present attendees.	✗	✗	✓	✓
A feature to share or upload notes efficiently	✗	✗	✓	✓
Safety Pocket Mode(feature to lock screen while in a meeting)	✗	✗	✗	✓
Effective use of a single link for students to join classes	✗	✗	✗	✓
A feature to send reminders for scheduled meetings	✗	✗	✓	✓
A limit present for a number of allowed attendees in the meeting	✗	✗	✓	✗
Highly secured with encryption	✓	✓	✓	✗
Reasonable pricing for all students	✗	✓	✓	✓

11.7 CONCLUSION

The e-learning platform has played a vital role in our lives, especially in the COVID-19 pandemic. These online tools supported us and made it easy to handle our worst situation globally. From now onwards, we are much more comfortable with online tools either in education or business. A detailed comparison is shown in the above table which comprises all the different features with proposed tool. Our work compares with other educational tools like Zoom, Microsoft Teams, Google Meet, and GoTo Meeting. The eScool online tool will be beneficial for the education sector and business domain as well. Two features are entirely different from other tools which are locking the screen while meeting and effective use of a single link for students to join classes but moderately secure with encryption.

Indeed, the popularity of online tools will increase in future in the education and business domains to connect worldwide. These meeting tools will make for a global village and provide us with a very useful platform. We can also solve all the complex and critical problems remotely. They allow us to conduct online classes, video conferencing, online interviews, business meetings, webinars, seminars, etc. These online applications have allowed us a sigh of relief. eScool will provide us with more features than other online applications which are available in the market.

REFERENCES

[1] Anand, Hardik. "Zoom Responds to MHA Deeming the App 'Not Safe'." *HT Tech*, Hindustan Times, 17 Apr. 2020, web.

[2] Chillcce, Angie Del Rio, et al. "Analysis of the Use of Videoconferencing in the Learning Process."

[3] "During the Pandemic at a University in Lima," *The Sai*, 5 Nov. 2021, web

[4] Perry, Tracy. "Overview of Security and Compliance – Microsoft Teams." Overview of Security and Compliance - Microsoft Teams | Microsoft Docs, Microsoft Docs, web.

[5] Rajput, Abhinav. "Zoom vs Google Meet vs Microsoft Teams: Which Video Conferencing App to Go For." HT Tech, Hindustan Times, 17 Apr. 2020, web.

[6] Singh, Ravinder, and Soumya Awasthi. "Updated Comparative Analysis on Video Conferencing Platforms- zoom, Google Meet, Microsoft Teams, WebEx Teams and Gotomeetings", 16 Aug. 2020, web.

[7] Soltero, Javier. "How Google Meet Is Helping Our G Suite and Google Cloud Customers Google Cloud Blog." Google, Google, 9 Apr. 2020, web.

[8] Ana Maria Suduc and Mihai Bizoi, AI shapes the future of web conferencing platforms "Comparing Zoom, Microsoft Teams and Google Meet." Devoteam G Cloud, 21 Dec. 2021, web.

[9] Zaghdoud, A., "Impact of Digital Transformation on Education Approaches: E-Learning – A Case Study of the National Office for Distance Education and Training," The University of Algiers, Algiers, Algeria, 2020.

[10] Englund, C., Olofsson, A. D., and Price, L. Teaching with technology in higher education: Understanding conceptual change and development in practice. *High. Educ. Res. Dev.* 36, 73–87, 2017. doi:10.1080/07294360.2016.1171300

[11] Saxena, K, *Coronavirus accelerates pace of digital education in India*. EDII Institutional Repository, 2020.

Chapter 12

Post-pandemic technology assisted teaching and learning
A perspective on self-directed learning

Shreya Virani[1] *and Sarika Sharma*[2]

[1]Symbiosis Centre for Management Studies, Symbiosis International (Deemed University), Pune, India

[2]Symbiosis Institute of Computer Studies and Research, Symbiosis International (Deemed University), Pune, India

12.1 INTRODUCTION

The global COVID-19 pandemic outbreak, which began in December 2019, affected almost all countries and territories. The pandemic has greatly hampered student prospects for global education. Students and teachers have the opportunity to set the foundation for the introduction of digital learning after being forced to transition from traditional classrooms to emergency online or remote learning [1]. As face-to-face instruction provided a way for online instruction, during this time period both students and educators had to deal with numerous difficulties. The COVID-19 pandemic has provided opportunities for both learners and teachers to get trained on various digital collaborative platforms [2]. As a result, they are now more confident about working with information and communications technology (ICT). Online learning, sometimes referred to as e-learning, synchronous learning, or asynchronous learning, is a novel method of education that uses cutting-edge communication technologies. E-learning platforms were developed to capitalize on the already-existing internet connectivity [3,4]. According to Khoza [5], the development of online learning environments (also known as learning management systems, or LMS) has rekindled interest in transforming conventional face-to-face teaching into a student-centred strategy characterized by self-directed learning (SDL). Thus, Self-Directed Learning (SDL) is a method by which students direct their own learning experience from start to end (Knowles, 1975). SDL is an imperative strategy for higher education and is essential to programmes for professionals as well [6,7]. Knowles [8,9] asserts that social interaction, including interactions with classmates, teachers, and tutors, is integral to learning. The learning spectrum can be thought of as having self-directed learning at one end and teacher-directed learning at the other [10].

Self-directed learning is a crucial component of a number of teaching and learning processes that students participate in higher education [11]. Self-directed learning requires freedom, which has become increasingly

DOI: 10.1201/9781003376699-12

important in the context of the digital revolution. Due to easier access to information technology, new channels of communication, and online learning communities, the meaning of lifelong learning has been expanded [12–14]. The user's ability to choose what, when, and how long to study has a huge impact on the efficiency of their learning efforts thanks to social media and other technology [15]. In order to understand the true dynamics of the interactions between learning and technologies, self-directed learning has been suggested as a viable and more direct route [16].

12.2 RATIONALE OF THE STUDY

The conceptions of teaching, learning, and instructor have changed, and new viewpoints and understandings have been added to these concepts, in order to better suit the demands of the current situation [17]. All aspects of life are immediately impacted by the world's rapid changes and revolutions in science and technology, as well as by their effects on educational procedures [18]. As a result of the COVID-19 epidemic, different applications in education have been deployed, including concepts such as blended learning and distance education.

Researchers interested in online learning are particularly interested in the student's aptitude for self-directed learning [19]. Students who learn how to be proactive and self-directed learners in college will be better equipped as workers to foresee the needs of their organizations. They may also acquire the required skills on their own and can create value for an organization as workers [20]. According to research, because of the COVID-19 situation, there was an acceleration in SDL to keep students engaged [21]. In this view, understanding the concept of self-directed learning from different perspectives has become more prevalent.

Earlier, the concept of SDL was more relevant to lifelong or distance learning programmes in higher education. With the recent changes in the education sector with the hit of the pandemic and the adoption of online teaching, the relevance of SDL has increased in the mainstream higher education programmes as well. The benefits of technology mediated learning are immense and therefore, even after the world has come back to traditional face-to-face teaching-learning, the scope for adoption of technology-enabled SDL is still there. This chapter explores the various aspects and dimensions of SDL and also investigates its feasibility and adaptability in higher educational institutions for a blended teaching-learning experience.

The present study provides a detailed review of literature on each element as per the selected definition, in order to gain a holistic perspective on the concept. SDL research studies have been undertaken for a while, but the context of online education and the extensive use of technology have not been explored. Specifically, in the context of the present pandemic situation, it is rather untouched. With the increased use of high-tech and sophisticated digital collaboration platforms available and used, research on the use of technology for SDL is much needed.

12.3 POST-PANDEMIC TEACHING-LEARNING AND SELF-DIRECTED LEARNING

Globally, the academic schedules of higher education institutions were disturbed by the COVID-19 pandemic. But every crisis also presents a chance for change. This paradigm might signify a move away. The service sectors where the core modus operandi was the physical and face-to-face interactions happened to be affected most. In developing countries like India, the education sector faces the challenges of survival. The educational institutions had to adopt digital technologies and had to switch to online teaching-learning to maintain continuity in education.

In post-digital learning communities, this necessitates a conceptual and philosophical reconsideration of the nature of teaching-learning activities, use of technology, infrastructure, and utilization of resources (Jandri et al., 2018). Many universities have been shifting to an online environment as a result of changing situations due to the pandemic. As a result of this, face-to-face deployment strategies for online learning and teaching have had to evolve [22]. The COVID-19 pandemic and online classes have led learners to engage themselves in self-directed learning through online learning [23]. The widespread consensus is that learners have greater control over their education when they learn online [24,25]. The addition of SDL to online learning increased students' freedom as they studied independently using the platform's resources. Self-directed learners spend more time preparing and evaluating learning outcomes, going through online resources and materials [26].

The transition to online training has accelerated self-directed learning, motivating and involving students in their learning more critically and independently. To enable self-directed learners to become active participants in their education rather than passive users of knowledge, a student-centred approach is essential [27]. As a result, self-directed learning is now required in order to improve student's proficiency. Students who had to become self-directed learners as a result of COVID-19 have experienced considerable effects [28]. With the implementation of COVID-19, SDL skills are regarded as being the most important for students [29]. Pan [30] looked at attitudes towards adopting technology for self-directed learning in research with undergraduate students from China. It has been demonstrated that self-directed learning and technological self-efficacy are related.

12.4 LITERATURE REVIEW

The authors have discussed the concept of self-directed learning as defined by Knowles (1975): "SDL is a process in which individuals actively participate, with or without the assistance of others, in identifying their learning needs, setting learning objectives, exploring human and material resources that are required for the learning purpose, selecting and implementing right learning strategies and assessing their learning outcomes."

12.4.1 An insight into the theoretical foundations of SDL

Self-directed learning is a multi-dimensional idea to be viewed from various perspectives. Van der Walt [31] also draws attention to the terminological ambiguity around this idea, which has complicated communication on the topic of self-directed learning. Furthermore, Dehnad et al. [32] claim that the definition of the term "self-directed learning" is inconsistent. Self-directed learning has many more definitions, whereas there is no consistency in theoretical perspectives (Candy, 1998). Draper [33] made an effort to connect self-directed learning to adult education by encouraging teachers to use alternative pedagogies for teaching and learning. With this background in mind, the current study examines some key conceptual frameworks that have been established by researchers in the context of self-directed learning to offer in-depth insight into the issue.

Self-directed learning is a strategy where students are encouraged to take on personal accountability and cooperative control of the cognitive (self-monitoring) and contextual (self-management) processes in order to construct and confirm meaningful and worthwhile learning outcomes, according to Garrison [34]. From the above discussion, it is evident that Garrison's concept of SDL emphasises the learner's motivation and collaborative perspective of the learner in order to accomplish the desired learning outcomes.

Grow's [35] *Staged Self-Directed Learning (SSDL) model* is among the most well-known self-directed learning methods. The SSDL model explains how learning can go from being reliant to being self-directed. The teaching strategies that are advised at various phases of a learner's self-direction are shown in Figure 12.1. Grow [35] suggested eight such factors to determine

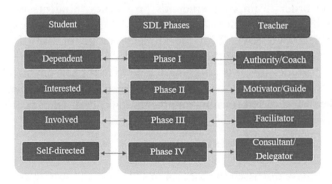

Figure 12.1 Staged self-directed learning model based on Grow [35] (own source).

one's readiness for self-directed learning as: "(1) openness to learning opportunities, (2) self-concept as an effective learner, (3) initiative and independence in learning, (4) informed acceptance of responsibility for one's own learning, (5) love of learning, (6) creativity, (7) positive orientation to the future, and (8) capacity for using fundamental study and problem-solving skills."

12.4.2 SDL is a process

Process is the "autonomous learning processes of learners." According to Moore [36,37], the act of organizing, overseeing, and evaluating one's own learning is the main way that learner autonomy is demonstrated. A six-step process can be used to describe self-directed learning: setting study goals, describing assessment in terms of how the learner will know when they are successful, identifying the structure and sequence of activities, setting up a schedule for completion of activities, identifying resources to help the learner reach each goal, and finding a mentor or faculty member to offer feedback on the plan [38]. In academic learning contexts, Winne and Hadwin [39] recognized four critical phases of self-directed learning: (1) identifying tasks; (2) Establishing goals and planning; (3) putting study methods and strategies into practice; and (4) metacognitive adapting studying. Self-directed learning is a process that involves social interaction, context, constructive feedback, self-regulation, and reflection [40,41]. In self-directed learning, the student takes charge of their learning objectives and strategies to achieve their own objectives or the needs they perceive to exist in their unique context. The fact that a learner's learning methods and goals are very customized and tailored to their particular circumstances in life is a distinguishing aspect of this process. The learners themselves represent a central and salient feature of their context.

Some researchers have investigated how online learning affects the SDL process and have focused on key areas like planning, monitoring, and evaluating in the context of SDL. Planning in online learning gives students the freedom to study at their own pace [42]. The flexibility of scheduling activities at a convenient time and location is a prominent feature of asynchronous online learning. Although the flexibility offered by online learning gives a student more freedom, it also comes with certain challenges (Song et al., 2004).

12.4.3 SDL is initiated by the individual

Learning is described psychologically, as the process through which the learners meet their needs and pursue their goals. This means that people are driven to learn when they have a personal goal they want to attain and a desire to learn. They are also motivated to use the resources that are accessible to them, such as teachers and reading material, when they believe such resources are pertinent to their needs and goals [43]. Individuals who self-direct their learning take the initiative and accountability for it. The idea that the student controls his or her own learning by taking ownership of and making decisions regarding what and how something is learnt is at the core of self-directed learning [44,45]. Self-directed learning can take place in both formal and informal settings [46,47]. According to Clardy [48], a person's former educational experiences are a reliable predictor of their predisposition to pursue self-directed learning. Self-directed learning behaviours can be developed in people when this is coupled with an environment that is encouraging and provides opportunity, resources, and dedication.

12.4.4 SDL may or may not involve the help of others

Garrison [49] stated that SDL in formal learning environments should be viewed as a process which facilitates between the teacher and the learner. Self-directed learning demands the ability to choose what to learn in addition to having the chance to do so. Therefore, formal education should be seen as a joint effort between the teacher and the learner. Merriam & Caffarella (2013) argue that there should be a greater acceptance of the inter-reliant and collaborative features of self-directed learning. O'Donnell [50] goes the farthest in accentuating the community over individual dimension when he explains the reason for what he calls "selves-directed learning." [8] asserts that social interaction, including interactions with peers, teachers, and tutors, is essential to learning. The best strategy teachers should think about using is to, encourage today's students to discover their own capacity for learning in order to reach the concept of "learn, unlearn, and relearn." [51]. Consequently, teachers and lecturers can aid in the development of self-directed learning [52].

However, SDL can be adopted inside as well as outside of the formal educational institution. Here researchers are keen on the SDL in educational institutions. Students need to develop and practise a variety of skills that allow them to direct their own learning rather than rely on a teacher to tell them what to do because SDL tends to be learner-centred rather than instructor-centred and instructor-driven. They must be capable of self-direction both individually and in peer groups, be self-conscious, aware of their value system, and possess qualities such as self-assurance and a positive self-concept [53,54].

12.4.5 Establishing learning objectives based on needs identification

Learning motivation, which is demonstrated by behaviour involvement in learning activities, is the process by which goal-directed action is initiated and sustained [55]. According to studies on students' learning, setting goals is closely associated with learning motivation [56–58]. The students should be very motivated to learn new things and accomplish their objectives. The personalized learning approach is addressed by SDL [59]. SDL is referred to as a goal that highlights a learner's desire or preference for taking ownership of their learning [60].

The self-directed learners have confidence in their capacity to learn. They can successfully make choices that are relevant to their learning requirements, and they view themselves gaining independence in this regard. Additionally, they are significantly more likely than their teacher-directed counterparts to feel successful as learners [46]. In their research study, Schweder and Raufelder [61] discovered that self-directed learning (SDL) encourages teenagers to satisfy their needs. In their study, Joo et al. [62] looked at how learning goal orientation, knowledge of developmental requirements, and self-directed learning affected employees' happiness with their careers in the Korean public sector. Based on an investigation using structural equation modelling, they found that learning goal orientation and awareness of developmental needs were accountable for the variability in self-directed learning. It has been demonstrated that self-directed learning (SDL) aids professionals in overcoming learning obstacles and gaining new abilities [63]. Additionally, evidence-based studies have demonstrated that digital learning supports SDL and that it is practical for professionals to attend courses through digital devices, favourably affecting learners' accomplishments [64–66].

12.4.6 Find the necessary resources to attain these goals

Resources come in different forms, including information resources and infrastructure. Students must be aware of and actively investigate a variety of learning tools [67]. When students are given access to resources, tools for learning, inspiration, and support, they can be helped to become more self-directed learners. As a result, a facilitator can serve as a counsellor, consultant, tutor, and resource finder in addition to being a classroom teacher [60]. Learning settings that effectively integrate technology may have a significant impact on independent learning because they give students access to materials in ways that were not previously possible [16]. In the aftermath of network technology, informal learning techniques such as online learning, e-learning, and others enable self-initiated design of learning experiences [68]. According to several research studies, self-directed learners get more from online learning than others do [69,70]. Students' perceptions of collaborative learning have the potential to improve their self-directed learning (SDL), according to research on SDLT [71]. The use of internet communication technologies for group learning is aided by student SDL procedures [72].

12.4.7 To achieve their objectives, choose and put into practice the best learning strategies

Effective learning strategies are to be adopted by learners for successful learning. SDL Model Learning strategies, according to Candy [73], describes how students approach a subject. Processing levels are divided into deep and surface levels. While surface learning is more concerned with reproducing the content, deep level processing seeks to find meaning in the subject. Deep-level processing complements study techniques like elaboration and searching for patterns and underlying ideas. Students who use a surface-level processing strategy are more likely to practise and memorize materials.

According to Khalid et al. [74], SDL and self-motivation go hand in hand. Learners believe they are active participants in their own education and have knowledge and skills. If they have a good guide, they are prepared to dive into a topic, and some do it alone. By using deliberate learning tactics, they will gain from knowing more about how they learn. The methodical aspect of SDL refers to the procedures (tasks and strategies) that students use to self-direct their learning. These procedures include learning by doing, cooperating, exhibiting, and discovering, as well as metacognition techniques (i.e., awareness of one's own mental processes, notably through reflection).

12.4.8 Determining how to measure learning outcomes

In SDL, learning has shifted its focus from teacher to student. This transition has facilitated students to choose their own goals and the steps to be taken to achieve their learning goals. Now the learning is student-centred. Peer evaluation contributes to better learning outcomes [75,76]. Self-assessment enables learners to diagnose their understanding through reflection in order to advance academically [77]. Self-assessment encompasses a variety of activities that students can engage in to evaluate their own performance, and this can be seen as a type of feedback and formative evaluation [78]. Boud [79] emphasises the value of self-evaluation in the educational process. Yan [80] establishes the value of self-assessment for self-regulated learning, implicitly reiterating its applicability to SDL. According to Garrison [34] self-directed learning encompasses certain processes for constructing learning outcomes.

12.4.9 Blended learning and SDL

A dynamic learning environment that accommodates multiple forms of communication is created through blended learning with combining in-person and digital learning, so that teaching and learning happens both in the classroom and online [81,82]. A mixed-learning course sits in the middle of a continuum with wholly online and fully face-to-face learning settings as its anchors [83]. Traditional learning, which includes face-to-face interaction and Information Technology (IT) components, must be successfully incorporated into the course design in order to ensure that blended learning is more than just an addition to the currently popular methodology or method (Kanuka, 2004). The kind of free and open conversation, critical debate, negotiation, and agreement that characterize good education are made possible in the setting of higher education through community engagement and interaction [84,85]. Blended learning is only successful and effective when students take ownership of and commitment to the learning (Bonk & Graham, 2006). According to Moore (2005), self-regulation and self-directed learning are related to blended learning because they involve three correlated elements: interaction, structure, and autonomy. By using an experimental research methodology Sriarunrasmee et al., [86] investigated the effectiveness of the blended learning approach on students' development of communication skills and self-directed learning. Their findings showed that students in blended learning classrooms outperformed students in traditional classrooms in both areas. Another crucial aspect of students' learning in the blended learning setting is their readiness for technology. The development of various computer technologies makes it possible to use multimedia content and multimedia communication in education (Horton, 2006) and makes learning content accessible at any time and from anywhere. Self-directed learners take an active role in their education and can adapt their learning methods to the circumstances. In a classroom with plenty of technology, students can tremendously benefit from the learning prospects and capabilities that are required to be self-directed in their learning [87].

In higher education, blended learning (BL), or the combination of in-person and online instruction, is extensively used. The phrase "new traditional model" [88] or "new normal" [89] has been used by certain academics to describe it. It has been mentioned that blended learning (BL) enables asynchronous teaching and learning by combining in-person and online training [90]. The online discussion forum is just one of the many online learning tools that BL uses to facilitate communication between students and between students and teachers. In contrast to the traditional classroom, blended learning enables educators and students to use all forms of communication, information technology, and innovation, particularly the Internet and networks, as tools to improve teaching and learning management as well as to produce educational media. For students to build their abilities and competence as learners, a blend of classroom and technology-assisted teaching-learning establishes access to a variety of modes and approaches to learning [91]. Numerous studies on blended learning demonstrate an improvement in students' capacity for collaborative learning, creative thinking, independent study, and personalization of their learning experiences. The use of blended learning by educators help them create avenues for communication that encourage students to share their knowledge and experiences [92]. The teachers' job in a blended learning setting is to act as a facilitator to help, suggest, and challenge students to meet their own learning objectives.

Asynchronous and collaborative learning among students is supported by an effective blending of traditional classroom instruction with online learning. However, students still enjoy face-to-face opportunities to receive feedback in BL settings, hence it is important to maintain a balance between traditional setting of classroom teaching, learning, and online learning [93].

12.5 DISCUSSIONS AND CONCLUSION

Technology enabled self-directed learning is here to stay and will affect the overall teaching-learning process in the long run. The stakeholders are keen to adopt SDL because of its obvious benefits offered. From the insights from previous sections, the researchers now propose a comprehensive model for technology enabled SDL based on Garrison [34] (Figure 12.2).

12.5.1 SDL from an educators' perspective

The educators can adopt SDL in their teaching with the help of blended learning activities. Along with the regular face-to-face teaching, some activities can also be planned to include SDL elements which can promote learner autonomy. With adequate technology support, teachers can inculcate more student-centric approaches and can enact more autonomous learning in the class. The long-term approaches of autonomy can be adopted in teaching practices.

12.5.2 SDL from learners' perspective

Higher education is a synonym for adult learning. The 21st century demands a variety of skills from learners. Certain areas which are important to developing knowledge are core content knowledge, digital literacy, problem

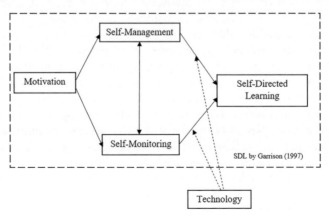

Figure 12.2 Dimensions of technology enabled self-directed Learning (own source).

solving, and critical thinking. According to Garison [34], the SDL helps in self-development, motivation, and self-monitoring. These three factors further support goal setting, interdependence, being reflective, critical thinking, and task motivation. Other than these, SDL when adopted with technology, also provides greater sources of information for the students.

12.5.3 SDL from a management perspective

Use of technology in teaching-learning leads to the optimum utilization of resources. The educational institutions may have to appoint fewer teachers if SDL is promoted, which will be helpful in situations where good and skilled teachers are hard to find. The availability of human resources will be optimized through technology-enabled SDL.

12.6 FUTURE DIRECTIONS

This study provides some meaningful insights into the technology that enables self-directed learning with respect to higher educational institutions. SDL for schools can also be explored further. It would be interesting to know in the future what factors may affect the SDL implementation at the institutes. Other than the benefits of SDL, are there any negative impacts on the student's learning can be another perspective for future research.

REFERENCES

[1] Dhawan, S. (2020). Online learning: A panacea in the time of COVID-19 crisis. *Journal of educational technology systems*, 49 (1), 5–22.

[2] Pokhrel, S., & Chhetri, R. (2021). A literature review on impact of COVID-19 pandemic on teaching and learning. *Higher Education for the Future*, 8 (1), 133–141.

[3] Mpungose, C. (2020). Emergent transition from face-to-face to online learning in a South African university in the context of the coronavirus pandemic. *Humanities and Social Sciences Communications*, 7 (1), 1–9. 10.1057/s41599-020-00603-x

[4] Ohlin, C. (2019). Information and communication technology in a global world. *Research in Social Sciences and Technology*, 4 (2), 41–57. 10.46303/ressat.04.02.4

[5] Khoza, S.B. (2019). Lecturers' reflections on curricular spider web concepts as transformation strategies. In *Transformation of higher education institutions in post-apartheid South Africa*. pp. 15–26, Routledge.

[6] Raidal, S.L., & Volet, S.E. (2009). Preclinical students' predispositions towards social forms of instruction and self-directed learning: a challenge for the development of autonomous and collaborative learners. *Higher Education*, 57 (5), 577–596.

[7] Sze-Yeng, F., & Hussain, R.M.R. (2010). Self-directed learning in a socio constructivist learning environment. *Procedia-Social and Behavioral Sciences, 9,* 913–1917.

[8] Knowles, M. (1990). *The adult learner: A neglected species.* Houston. TX: Gulf Publishing.

[9] Lai, C., & Gu, M.Y. (2011). Self-regulated out-of-class language learning with technology. *Computer Assisted Language Learning, 24* (4), 317–335. doi: 10.1080/09588221.2011.568417

[10] Fisher, M., King, J., & Tague, G. (2001). Development of a self-directed learning readiness scale for nursing education. *Nurse Education Today, 22* (7), 516–525.

[11] Rascón-Hernán, C., Fullana-Noell, J., Fuentes-Pumarola, C., Romero-Collado, A., Vila-Vidal, D., & Ballester-Ferrando, D. (2019). Measuring self-directed learning readiness in health science undergraduates: A cross-sectional study. *Nurse Education Today, 83,* 104201.

[12] Kim, R.H. (2010). *Self-directed learning management system: Enabling competency and self-efficacy in online learning environments.* The Claremont Graduate University.

[13] Thorpe, M. (2005). The impact of ICT on lifelong learning, *Lifelong learning & distance higher,* pp. 23–32. Retreived from: https://universidadazteca.net/yahoo_site_admin/assets/docs/141218e_Lifelong_Learning_Distance_HE.63165716.pdf#page=32 (accessed on 20 March, 2023).

[14] Tobin, D.R. (2000). *All learning is self-directed: How organizations can support and encourage independent learning.* Alexandria, VA: American Society for Training and Development.

[15] Tullis, J.G., & Benjamin, A.S. (2011). On the effectiveness of self-paced learning, *Journal of memory and language, 64* (2), 109–118.

[16] Candy, P.C. (2004). *Linking thinking: Self-directed learning in the digital age,* Canberra, Australia, Department of Education, Science and Training.

[17] Karataş, K., Şentürk, C., & Teke, A. (2021). The mediating role of self-directed learning readiness in the relationship between teaching-learning conceptions and lifelong learning tendencies. *Australian Journal of Teacher Education, 46* (6), 54–77. 10.14221/ajte.2021v46n6.4

[18] UNCTAD (2018). *The impact of rapid technological change on sustainable development.* United Nations Publishing.

[19] Hartley, K., & Bendixen, L.D. (2001). Educational research in the Internet age: Examining the role of individual characteristics. *Educational Researcher, 30* (9), 22–26.

[20] Artis, A.B., & Harris, E.G. (2007). Self-directed learning and sales force performance: An integrated framework. *Journal of Personal Selling and Sales Management, 27* (1), 9–24.

[21] Adinda, D., & Mohib, N. (2020). Teaching and instructional design approaches to enhance students' self-directed learning in blended learning environments. *The Electronic Journal of e-Learning, 18*(2),162–174. 10.34190/EJEL.20.18.2.005.

[22] Owolabi, J.O. (2020). Virtualising the school during COVID-19 and beyond in Africa: infrastructure, pedagogy, resources, assessment, quality assurance, student support system, technology, culture and best practices. *Advances in Medical Education and Practice. 11,* 755–759. doi: 10.2147/AMEP.S272205

[23] Baticulon, R.E., Sy, J.J., Alberto, N.R.I., Mabulay, R.E.C., Rizada, L.G.T., Tiu, C.J.S., Clarion, C.A., & Reyes. J.C.B. (2021). Barriers to online learning in the time of COVID-19: A national survey of medical students in the Philippines. *Medicine Science Education*, 31, 615–626. 10.1007/s40670-021-01231-z

[24] Garrison, D.R. (2003). Cognitive presence for effective asynchronous online learning: The role of reflective inquiry, self-direction and metacognition. *Elements of quality online education: Practice and direction*, 4 (1), 47–58.

[25] Gunawardena, C.N., & McIsaac, M.S. (2003). Distance education. D. Jonassen (Ed.), *Handbook for research on educational communications and technology* (pp. 355–396). New York.

[26] Geng, S., Law, K.S. & Niu, B. (2019). Investigating self-directed learning and technology readiness in blending learning environment. *International Journal of Educational Technology in Higher Education*, 16 (1), 1–22. 10.1186/s41239-019-0147-0

[27] Roberson Jr., D.N., Zach, S., Choresh, N., & Rosenthal, I. (2021). Self-directed learning: A longstanding tool for uncertain times. *Creative Education*, 12 (5), 1011–1026. 10.4236/ce.2021.125074

[28] Mukhtar, K., Javed, K., Arooj, M., & Sethi, A. (2020). Advantages, limitations and recommendations for online learning during COVID-19 pandemic era. *Pakistan Journal of Medical Sciences*, 36, 21–27.

[29] Morris, T.H. (2021). Meeting educational challenges of pre-and post-COVID-19 conditions through self-directed learning: Considering the contextual quality of educational experience necessary. *On the Horizon*, 29 (2), 1–20. doi:10.1108/OTH01-2021-0031.

[30] Pan, X. (2020). Technology acceptance, technological self-efficacy, and attitude toward technology-based self-directed learning: Learning motivation as a mediator. *Frontiers in Psychology*, 11, 1–11. 10.3389/fpsyg.2020.564294.

[31] Van der Walt, J.L. (2019). The term Self-Directed Learning - back to Knowles, or another way to forge ahead? *Journal of Research on Christian Education*, 28 (1), 1–20.

[32] Dehnad, A., Afsharian, F., Hosseini, F., Arabshahi, S.K.S., & Bigdeli, S. (2014). Pursuing a definition of self-directed learning in literature from 2000–2012. *Procedia-Social and Behavioral Sciences*. 116, 5184–5187.

[33] Draper, J. (1998). The metamorphoses of andragogy. *Canadian Journal for the Study of Adult Education*, 12 (1), 3–26.

[34] Garrison, D.R. (1997). Self-directed learning: Toward a comprehensive model, *Adult Education Quarterly*, 48 (1), 8–33. 10.1177/074171369704800103.

[35] Grow, G.O. (1991). Teaching learners to be self-directed, *Adult Education Quarterly*, 41 (3), 125–149.

[36] Moore, M.G. (1972). Learner autonomy: The second dimension of independent learning. *Convergence*, 5 (2), 76–88.

[37] Moore, M., & Kearsley, G. (2005). *Distance education: A system view*. Boston: Wadsworth.

[38] Robinson, J.D., & Persky, A.M. (2020). Developing self-directed learners. *American Journal of Pharmaceutical Education*, 84 (3), 847512. 10.5688/ajpe847512.

[39] Winne, P.H., & Hadwin, A.F. (1998). Studying as self-regulated engagement in learning. In D. Hacker, J. Dunlosky, & A. Graesser (Eds.), *Metacognition in educational theory and practice*, pp. 277–304, Hillsdale: Lawrence Erlbaum.

[40] Simons, P.R.J. (2000). Towards a constructivistic theory of self-directed learning. *Self-learning*, pp. 1–12 retrieved from file:///C:/Users/viran/Downloads/5701%20(2).pdf (accessed on 20 March 2023).

[41] Song, L. (2005). *Self-directed learning in online environments: process, personal attribute, and context* (Doctoral dissertation, University of Georgia).

[42] Chizmar, J.F., & Walbert, M.S. (1999). Web-based learning environments guided by principles of good teaching practice. *Journal of Economic Education, 30 (3)*, 248–264.

[43] Chene, A. (1983). The concept of autonomy in adult education: A philosophical discussion. *Adult Education Quarterly, 34 (1)*, 38–47.

[44] Merriam, S.B., & Bierema, L.L. (2013). *Adult learning: Linking theory and practice*. New Jersey: Jossey-Bass, John Wiley & Sons.

[45] Mishra, P., Fahnoe, C., Henriksen, D., & Deep-Play Research Group (2013). Creativity, self-directed learning and the architecture of technology rich environments. *TechTrends, 57 (1)*, 10–13.

[46] Loeng, S. (2020). Self-directed learning: A core concept in adult education. *Education Research International, 2020*, 1–20.

[47] Means, B., Toyama, Y., Murphy, R., Bakia, M., & Jones, K. (2009). Evaluation of evidence-based practices in online learning: A meta-analysis and review of online learning studies, available at https://repository.alt.ac.uk/629/1/US_DepEdu_Final_report_2009.pdf (accessed on 10 December, 2022).

[48] Clardy, A. (2000). Learning on their own: Vocationally oriented self-directed learning projects. *Human Resource Development Quarterly, 11 (2)*, 105–125.

[49] Garrison, D.R. (1992). Critical thinking and self-directed learning in adult education: an analysis of responsibility and control issues, *Adult Education Quarterly, 42 (3)*, 136–148.

[50] O'Donnell, D. (1999). Habermas, critical theory and selves-directed learning. *Journal of European Industrial Training, 23 (4/5)*, 251–261.

[51] Piseth, N. (2020). Self-directed learning: The way forward for education after the COVID-19 crisis. Available at https://cefcambodia.com/2020/08/03/self-directed-learning-the-way-forward-for-education-after-the-covid-19-crisis/ (accessed on 10 February, 2023)

[52] Cadorin, L., Bortoluzzi, G., & Palese, A. (2013). The self-rating scale of self-directed learning (SRSSDL): A factor analysis of the Italian version, *Nurse Education Today, 33 (12)*,1511–1516.

[53] Giveh, F. (2018). SDL via contemplative teaching to promote reading comprehension ability. *English Language Teaching, 11(12)*, 58–76.

[54] Shahrory, E. (2013). The degree of SDL skills obtains for university level students in Riyadh. *Journal of Educational Sciences Studies, 40*, 927–944.

[55] Fredricks, J.A., Blumenfeld, P.C., & Paris, A.H. (2004). School engagement: Potential of the concept, state of the evidence. *Review of Educational Research, 74 (1)*, 59–109.

[56] Che-Ha, N., Mavondo, F.T., & Mohd-Said, S. (2014). Performance or learning goal orientation: Implications for business performance. *Journal of Business Research*, 67 (1), 2811–2820.
[57] Law, K.M., & Breznik, K. (2017). Impacts of innovativeness and attitude on entrepreneurial intention: Among engineering and non-engineering students. *International Journal of Technology and Design Education*, 27 (4), 683–700.
[58] Law, K.M., Lee, V.C., & Yu, Y.T. (2010). Learning motivation in e-learning facilitated computer programming courses. *Computers & Education*, 55 (1), 218–228.
[59] Lalitha, T.B., & Sreeja, P.S. (2020). Personalised self-directed learning recommendation system. *Procedia Computer Science*, 171, 583–592.
[60] Brockett, R.G. & Hiemstra, R. (1991). *Self-direction in adult learning: Perspectives on theory, research, and practice*, New York, NY: Routledge.
[61] Schweder, S., & Raufelder, D. (2021). Needs satisfaction and motivation among adolescent boys and girls during self-directed learning intervention. *Journal of Adolescence*, 88, 1–13.
[62] Joo, B.K., Park, S., & Oh, J.R. (2013). The effects of learning goal orientation, developmental needs awareness and self-directed learning on career satisfaction in the Korean public sector. *Human Resource Development International*, 16 (3), 313–329.
[63] Jeong, D., Presseau, J., ElChamaa, R., Naumann, D.N., Mascaro, C., Luconi, F., Smith, K.M. & Kitto, S. (2018). Barriers and facilitators to self-directed learning in continuing professional development for physicians in Canada: a scoping review. *Academic Medicine*, 93 (8), 1245–1254.
[64] Berndt, A., Murray, C.M., Kennedy, K., Stanley, M.J., & Gilbert-Hunt, S. (2017). Effectiveness of distance learning strategies for continuing professional development (CPD) for rural allied health practitioners: a systematic review. *BMC Medical Education*, 17 (1), 1–13.
[65] Bonk, C.J., & Graham. C.R., (Eds.). (2004). *Handbook of blended learning: Global perspectives, local designs*, San Francisco, CA: Pfeiffer Publishing.
[66] Scott, K.M., Baur, L., & Barrett, J. (2017). Evidence-based principles for using technology-enhanced learning in the continuing professional development of health professionals. *Journal of Continuing Education in the Health Professions*, 37 (1), 61–66.
[67] Sener, J., & Stover, M.L. (2000). Integrating ALN into an independent study distance education program: NVCC case studies. *Journal of Asynchronous Learning Networks*, 4 (2), 126–144.
[68] Reinders, H., & White, C. (2011). Learner autonomy and new learning environments. *Language Learning & Technology*, 15 (3), 1–3.
[69] Lee, J., Hong, N.L., & Ling, N.L. (2002). An analysis of students' preparation for the virtual learning environment. *Internet and Higher Education*, 5 (3), 231–242.
[70] Shapley, P. (2000). On-line education to develop complex reasoning skills in organic chemistry, *Journal of Asynchronous Learning Networks*, 4, 43–49.
[71] Teo, T., Tan, S.C., Lee, C.B., Chai, C.S., Koh, J.H.L., Chen, W.L., & Cheah, H.M. (2010). The self-directed learning with technology scale (SDLTS) for young students: An initial development and validation, *Computers & Education*, 55 (4),1764–1771. 10.1016/j.compedu.2010.08.001

[72] Lee, K., Tsai, P.S., Chai, C.S., & Koh, J.H.L. (2014). Students' perceptions of self-directed learning and collaborative learning with and without technology. *Journal of Computer Assisted Learning, 30* (5), 425–437. 10.1111/jcal.12055

[73] Candy, P.C. (1991). Self-Direction for Lifelong Learning. A Comprehensive Guide to Theory and Practice. Jossey-Bass, 350 Sansome Street, San Francisco, CA 94104-1310.

[74] Khalid, M., Bashir, S., & Amin, H. (2020). Relationship between Self-Directed Learning (SDL) and academic achievement of university students: A case of online distance learning and traditional universities. *Bulletin of Education and Research, 42* (2), 131–148.

[75] Li, H., Xiong, Y., Hunter, C.V., Guo, X., & Tywoniw, R. (2020). Does peer assessment promote student learning? A meta-analysis. *Assessment & Evaluation in Higher Education, 45* (2), 193–211. 10.1080/02602938.2019.1620679

[76] Sanchez, C.E., Atkinson, K.M., Koenka, A.C., Moshontz, H. & Cooper, H. (2017). Self-grading and peer-grading for formative and summative assessments in 3rd through 12th grade classrooms: A meta-analysis. *Journal of Educational Psychology, 109* (8), 1049–1066.

[77] Baez, Z.D. (2019). ICT and its purpose in the pedagogical practice. *Research in Social Sciences and Technology, 4* (2), 83–95. 10.46303/ressat.04.02.6

[78] Andrade, H.L. (2019). A critical review of research on student self-assessment. *Frontiers in Education, 4,* 1–13. 10.3389/feduc.2019.00087

[79] Boud, D. (2013). *Enhancing learning through self-assessment,* New York, NY: Routledge.

[80] Yan, Z. (2020). Self-assessment in the process of self-regulated learning and its relationship with academic achievement. *Assessment & Evaluation in Higher Education, 45* (2), 224–238.

[81] Collis, B., & Moonen, J. (2012). *Flexible learning in a digital world: Experiences and expectations,* London and New York: Routledge, Taylor & Francis Group.

[82] Cron, W.L., Marshall, G.W., Singh, J., Spiro, R.L., & Sujan, H. (2005). Salesperson selection, training, and development: Trends, implications, and research opportunities. *Journal of Personal Selling and Sales Management, 25* (2), 123–136.

[83] Rovai, A.P., & Jordan, H.M. (2004). Blended learning and sense of community: A comparative analysis with traditional and fully online graduate courses. *International Review of Research in Open and Distributed Learning, 5* (2), 1–13. 10.19173/irrodl.v5i2.192

[84] Garrison, D.R., & Cleveland-Innes, M. (2005). Facilitating cognitive presence in online learning: Interaction is not enough. *The American Journal of Distance Education, 19* (3), 133–148.

[85] Garrison, R., & Kanuka, H. (2004). Blended learning: Uncovering its transformative potential in higher education. *Internet and Higher Education, 7* (2), 95–105.

[86] Sriarunrasmee, J., Techataweewan, W., & Mebusaya, R.P. (2015). Blended learning supporting self-directed learning and communication skills of Srinakharinwirot University's first year students. *Procedia-Social and Behavioral Sciences, 197,* 1564–1569.

[87] Fahnoe, C., & Mishra, P. (2013). Do 21st century learning environments support self-directed learning? Middle school students' response to an intentionally designed learning environment. In *Society for information technology & teacher education international conference* (3131–3139). Association for the Advancement of Computing in Education (AACE).

[88] Ross, B., & Gage, K. (2006). *Global perspectives on blending learning* (pp. 155–168). San Francisco: Pfeiffer.

[89] Norberg, A., Dziuban, C.D., & Moskal, P.D. (2011). A time-based blended learning model. *On the Horizon, 19* (3), 207–216. 10.1108/10748121111163913

[90] Graham, C.R. (2013). Emerging practice and research in blended learning. In M.G. Moore (Ed.), *Handbook of distance education*, 333–350, New York, NY: Routledge.

[91] Cleveland-Innes, M., Ostashewski, N., & Wilton, D. (2017). iMOOCs and learning to learn online. *Community of Inquiry Blog Post*, Retrieved from: https://www.thecommunityofinquiry.org/project5 (Accessed on 13 October 2022).

[92] Orhan, R. (2008). Redesigning a course for blended learning environment. *Turkish Online Journal of Distance Education, 9* (1), 54–66.

[93] Vanslambrouck, S., Zhu, C., Lombaerts, K., Philipsen, B., & Tondeur, J. (2018). Students' motivation and subjective task value of participating in online and blended learning environments. *The Internet and Higher Education, 36*, 33–40.

Chapter 13

Education 5.0

An overview

B V Babu

Consultant on Quality Assurance in Higher Education, India

13.1 INTRODUCTION

Quality Assurance in Higher Education is an essential and integral part of various assessment and accreditation bodies and rankings across the world. To make a mark in the academic world, to get the required recognition, and also to be in the race, obtaining these accreditations and rankings will be of utmost importance.

To ensure the assured quality in the outcomes of higher education, the following aspects are important and hence the understanding of these aspects is inevitable:

1. Value-based education
2. Research-based learning
3. Project-based learning
4. Experiential learning
5. Student aspirations
6. Flexibilities
7. Industry 5.0
8. Curriculum design
9. Teaching-learning-evaluation processes
10. Outcome-based education

There has been a paradigm shift from the conventional one-way monotonous teacher-centric learning to two-way experiential student-centric learning. Education 5.0 encompasses this paradigm shift in learning process and all the ten aspects as mentioned above. Education 5.0 establishes a seamless integration of all these aspects and components.

Education 5.0 represents a new concept in education that emphasises on meeting the needs of the 21st-century learner [1,2]. By emphasizing personalization, lifelong learning, emerging technologies, global collaboration, and skills and competencies, Education 5.0 is designed to prepare learners to be successful in today's rapidly changing world. It focuses on skills and competencies that are essential for success in today's continuously changing world.

Education 5.0 is a term used to describe the latest paradigm shift in education that is being driven by the rapid advances in technology and globalization. It represents a novel approach to education that ensures on developing students who are adaptable, innovative, and equipped with the required skills and knowledge base to survive in this ever changing world.

Education 5.0 builds upon the previous generations of education, which were characterized by a focus on knowledge acquisition (Education 1.0), mass education and standardization (Education 2.0), personalized learning and competency-based education (Education 3.0), and digital and connected learning (Education 4.0). Education 5.0 takes a more holistic and student-centric approach to education, and places a greater emphasis on experiential learning, global citizenship, values-based education, and enhancement of skills such as critical thinking abilities, problem-solving capabilities, creativity, and collaborative work.

Through Education 5.0, the emphasis is on developing lifelong learners who can learn and adapt to new situations throughout their life span. This is achieved through the use of personalized and adaptive learning technologies, which enable students to learn in their way and at the pace they want. Education 5.0 also emphasizes the importance of global citizenship and cultural understanding, with an emphasis on collaboration and cooperation across borders and cultures.

Overall, Education 5.0 represents a significant shift in education that recognizes the importance of making students ready for the ever changing and interconnected world. By focusing on developing students who are adaptable, innovative, and equipped with the required skills and essential knowledge base to survive in this competitive world, Education 5.0 has the potential to transform education and empower individuals to achieve their full potential.

In this chapter, we shall first discuss in detail the above quality assurance aspects and then further explore how seamlessly these aspects along with student-centric learning have been integrated into Education 5.0.

In Part I of this chapter, a brief introduction is given (as above) on Education 5.0. Part II of this chapter elaborates on the Quality assurance aspects and Student-centric learning. Part III of this chapter gives a detailed account of Education 5.0, covering and seamlessly integrating all the aspects associated with it. Part IV of this chapter summarises the challenges associated with Education 5.0 and provides possible solutions to address them. Finally, the conclusions on Education 5.0 are presented in Part V of this chapter. References are listed at the end as the last section of this chapter.

13.2 QUALITY ASSURANCE ASPECTS & STUDENT-CENTRIC LEARNING

13.2.1 Value-based education

Value-based education is one of the most important approaches to education, which emphasizes the development of ethical and moral values in learners. The goal of value-based education is to instil in learners a responsibility, empathy, and compassion, and to prepare them to be ethical and responsible members of society [3].

It includes the understanding of the inherent connect among self, society, and nature. With the advent of scientific breakthroughs and technological discoveries, inventions, and innovations over the past many centuries, the human beings somewhere lost their humility, sensitivity, and sensibilities to other beings in nature [4,5]. Now, the time has come to explore within to understand these values and hence the value-based education plays an important role at all three levels of education (primary, secondary, and higher education spheres), more so at higher education level [6,7].

The educators should train the students to be aware of the connect among the self, society, and nature and ensure that this connect does not get affected during their scientific discoveries and technological inventions. The connect can be ensured by not interfering with four orders of nature, five dimensions of human order, and nine feelings in relationships among them in any of their scientific and technological discoveries and inventions [8,9].

The four orders of nature include "(1) Bio order (plants, trees, etc.), (2) Physical order (soil, metals, etc.), (3) Animal order (animals, birds, etc.), and (4) Human order (human beings)'.

The five dimensions of human order consist of '(1) Education (Sanskar), (2) Health (self-regulation), (3) Production (work), (4) Justice (preservation), and (5) Exchange (storage)'.

The nine feelings of relationship encompass '(1) Trust (viswas, the foundation value), (2) Respect (samman), (3) Affection (sneh), (4) Care (mamta), (5) Guidance (vatsalya), (6) Reverence (shraddha), (7) Glory (gaurav), (8) Gratitude (krutaznatha) and (9) Love (prem, the complete value)' [10].

Value-based education is characterized by the features given below:

1. Values-based curriculum: Value-based education emphasizes a curriculum that is designed to teach learners the values of honesty, respect, kindness, empathy, responsibility, and compassion.
2. Character education: Value-based education focuses on character education, where learners are taught to have positive character traits, viz., honesty, integrity, and empathy.
3. Service learning: Value-based education promotes service learning, where learners engage in community service projects that help to develop a sense of social responsibility and civic engagement.

4. Positive school culture: Value-based education promotes a positive school culture, where learners are encouraged to demonstrate respect, kindness, and empathy towards others.
5. Role models: Value-based education emphasizes the importance of role models, where learners are exposed to positive role models who exemplify the values and traits that are being taught.

Value-based learning emphasizes the ethical and moral values. It is designed to prepare learners to be responsible and ethical members of society, and to develop a sense of social responsibility and civic engagement [11].

13.2.2 Research-based learning

Research-based learning approach in education is all about the use of research in teaching and learning practices. The goals of research-based learning is to use the latest research findings to improve the quality aspects of teaching and learning, and to ensure that educational practices are based on sound evidence.

The features of research-based learning typically are [12]

1. Evidence-based teaching practices: Research-based learning focuses on the use of evidence-based teaching practices, where teaching strategies are based on the latest research findings on effective practices of teaching and learning strategies.
2. Action research: Research-based learning promotes the use of action research, where educators engage in systematic inquiry in improving their teaching strategies and to evaluate the effectiveness of educational programs.
3. Data-driven decision making: Research-based learning emphasizes the use of data to make informed decisions, where educational practices are evaluated based on data-driven evidence of their effectiveness.
4. Continuous improvement: Research-based learning emphasizes the importance of continuous improvement, where educational practices are evaluated and refined based on ongoing research and evaluation.
5. Collaboration and knowledge-sharing: Research-based learning promotes collaboration and knowledge-sharing among educators, researchers, and practitioners, to ensure that the latest research findings are disseminated and applied in educational settings.

Hence, research-based learning is designed to basically ensure that educational practices are based on sound evidence, and to promote continuous improvement and innovation in teaching and learning.

13.2.3 Project-based learning

Project-based education in teaching and learning process emphasizes hands-on, experiential learning through the completion of projects or tasks. In project-based education, learners work collaboratively to identify and solve real-world problems, create products or services, and develop skills and competencies through a process of inquiry and exploration [13].

Here are some common features of project-based learning:

1. Student-centered learning: Project-based learning is student-centered, where learners play an active role in their learning process by identifying and pursuing their own interests and goals.
2. Authentic tasks: Project-based learning involves the completion of authentic tasks that are relevant and meaningful to learners' lives and interests, and that have real-world applications.
3. Collaborative learning: Project-based learning emphasizes collaborative learning, through which learners work together by sharing ideas, solving problems, and creating resultant products or services.
4. Inquiry-based learning: Project-based learning involves inquiry-based learning, where learners engage themselves in the process of exploration and discovery, and develop the innate skills such as critical thinking and problem-solving.
5. Assessment: Project-based learning involves ongoing assessment, where learners receive feedback on their progress and use it to improve their skills and competencies.

Project-based learning is designed to engage learners in their learning process and focuses on developing skills and competencies that are relevant and meaningful to their lives and interests. By providing opportunities for inquiry, collaboration, and authentic learning experiences, project-based education can help learners to develop the skills and competencies that are needed to succeed in the fast changing world.

13.2.4 Experiential learning

Experiential learning focuses on emphasizing the importance of hands-on real-world experiences among the students in their learning process. Learners are encouraged to apply their acquired knowledge and skills to real practical situations, and develop a deeper understanding of the world around them [14,15].

Experiential learning emphasizes and typically characterized by the following features:

1. Active participation: Through active participation, learners play an active role in their learning process through activities such as internships, apprenticeships, and project-based learning.

2. Reflection: Experiential learning focuses on the importance of reflection, where learners reflect on their experiences and identify what they have learned.
3. Application: Explains the importance of applying knowledge and skills to practical situations, where learners can see the relevance and applicability of what they have learned.
4. Feedback: Experiential learning also emphasizes the importance of feedback, where learners receive feedback continuously on their performance and enable them to use it to improve their skills and competencies.

Experiential learning is an approach to learning that emphasizes the importance of hands-on, real-world experiences in the learning process. It is basically designed to provide learners with opportunities to apply their acquired knowledge and skills to real-world practical situations and enable them to develop a deeper understanding of the world around them. Experiential learning is often used in vocational and technical education, but it can be applied to any subject or field of study.

13.2.5 Student aspirations

Students are the main stakeholders in an education system. However, the students aspirations by far have been the most neglected aspect in curriculum design and framework in higher education. Irrespective of the degree one obtains, the students could be broadly categorized into six groups, based on their aspirations, such as (1) gaining good social stature by obtaining a good job in hand before passing out of the university (Campus placements), (2) becoming a job provider rather than a job seeker (Entrepreneurship), (3) becoming an academician (pursuing higher studies abroad), (4) becoming a good researcher (project-based and research-based learning, (5) becoming a good performing arts professional (Music, Dance, Dramatics, Photography, Film making, Spirituality, etc.), and (6) Administrative services through competitive examinations for public service commission and probationary officers, etc. [16].

Student aspirations refer to the goals, ambitions, and desires that students have for their future. Aspirations can be related to academic, personal, and professional goals, and they would play a significant role in motivating students to realize their full potential.

Common types of student aspirations:

1. Academic aspirations: Many students aspire to achieve academic success, such as getting good grades, graduating with honors, and pursuing further education.
2. Career aspirations: Many students aspire to pursue a specific career path or profession, such as becoming a doctor, lawyer, engineer, or entrepreneur.

3. Personal aspirations: Some students aspire to develop personal qualities and skills, such as becoming more confident, improving communication skills, or becoming more socially engaged.
4. Creative aspirations: A few students aspire to develop their creative talents, such as music, art, or writing.
5. Social aspirations: Some students aspire to be able to make a positive impact on their local communities or world at large through social activism, community service, or environmental advocacy.
6. Entrepreneurial aspirations: These refer to the desire to start and run a business venture, and these aspirations have become an important component of student aspirations in recent years.

13.2.5.1 Academic aspirations

Academic aspirations refer to the educational goals and ambitions of students, which may include pursuing higher education, achieving academic excellence, and acquiring specialized knowledge and skills in a particular field. Academic aspirations are shaped by a variety of factors, such as family background, cultural values, personal interests, and career aspirations [17].

Some common features of academic aspirations:

1. Higher education: Many students aspire to pursue higher education (obtaining a college degree or a graduate degree), to gain specialized knowledge and skills in a particular field and increase their career prospects.
2. Academic excellence: Many students aspire to achieve academic excellence, by earning high grades, winning academic awards, and pursuing challenging courses and projects.
3. Specialized knowledge and skills: Some students aspire to acquire specialized knowledge and skills in a particular field of their interest, viz., science, technology, engineering, and mathematics (STEM) or the arts and humanities.
4. Career aspirations: Some students aspire to pursue careers in specific fields, such as medicine, law, business, or technology, and may tailor their academic aspirations to achieve their career goals.
5. Personal growth: Academic aspirations may also include personal growth and development, such as acquiring new perspectives, developing critical thinking skills, and pursuing intellectual curiosity.

Academic aspirations play an important role in shaping students' educational journeys and career trajectories. By setting clear goals and aspirations, students can focus their efforts and energies on achieving their desired outcomes and realizing their full potential [18].

13.2.5.2 Career aspirations

Career aspirations refer to the career goals and ambitions of individuals, which may include pursuing a particular profession, acquiring specific skills and experience, or achieving a certain level of success in their chosen career path. Career aspirations get realised by a variety of factors based on their personal interests, values, personality traits, and work experiences [19].

Common features of career aspirations:

1. Career choice: Career aspirations often involve choosing a specific career path, such as becoming a doctor, lawyer, engineer, or artist.
2. Skill development: Career aspirations may involve acquiring specific skills and experience in a particular field, such as leadership skills, technical expertise, or project management skills.
3. Advancement: Career aspirations may involve advancing within a chosen career path, such as moving from an entry-level position to a management position, or starting a business.
4. Work-life balance: Career aspirations may also involve achieving a balance between work and personal life, such as working in a flexible work environment or pursuing a career that aligns with personal values and interests.
5. Personal fulfilment: Career aspirations may also involve achieving personal fulfilment and satisfaction through one's work, such as making a positive impact on society, pursuing a passion, or achieving financial stability.

Career aspirations are an important part of career planning and can help individuals achieve their professional and personal goals. By setting clear career goals and developing a plan to achieve them, individuals can make informed decisions about their education, work experiences, and skill development, and increase their chances of success in their chosen career path [20].

13.2.5.3 Personal aspirations

Personal aspirations refer to the personal goals and ambitions of individuals, which may include achieving personal growth, pursuing new experiences, building meaningful relationships, or finding inner peace and fulfilment. Personal aspirations are imbibed by a variety of factors including personal values, life experiences, interests, and relationships [21].

Some features of personal aspirations:

1. Personal growth: Many individuals aspire to achieve personal growth and development, such as developing new skills, gaining new knowledge, and broadening their perspectives.

2. New experiences: Many individuals aspire to explore new experiences and activities, such as traveling to new places, trying new hobbies, or taking on new challenges.
3. Meaningful relationships: Many individuals aspire to build meaningful relationships with family, friends, or romantic partners, and to cultivate a sense of belonging and connection.
4. Inner peace and fulfilment: Many individuals aspire to find inner peace and fulfilment, by exploring their spirituality, practicing mindfulness, or pursuing a life purpose.
5. Health and well-being: Many individuals aspire to achieve optimal health and well-being, by adopting healthy habits, staying physically active, and prioritizing self-care.

Overall, personal aspirations are an important part of self-discovery and personal growth, and can help individuals live a more fulfilling and satisfying life. By setting clear personal goals and working towards achieving them, individuals can develop a deeper understanding of themselves, build meaningful relationships, and find greater happiness and fulfilment [22].

13.2.5.4 Creative aspirations

Creative aspirations refer to the creative goals and ambitions of individuals, which may include pursuing artistic endeavors, developing new ideas and innovations, or exploring new modes of self-expression. Creative aspirations are shaped up by many factors, such as personal interests, life experiences, and exposure to different forms of art and culture [23].

Common features of creative aspirations [23]:

1. Artistic expression: Many individuals aspire to express themselves creatively through various forms of art, such as painting, writing, music, dance, or theater.
2. Innovation: Many individuals aspire to develop new ideas and innovations that push the boundaries of traditional art and culture, such as creating new genres of music or developing new visual art forms.
3. Self-expression: Many individuals aspire to use their creativity as a means of self-expression, to convey their thoughts, feelings, and experiences in a unique and meaningful way.
4. Collaboration: Many individuals aspire to collaborate with others in creative endeavors, to explore new perspectives, and create new forms of art and culture.
5. Impact: Many individuals aspire to use their creativity as a means of making a positive impact on society, by raising awareness of social issues, promoting social justice, or inspiring change.

Creative aspirations are an important part of human expression and can help individuals develop a deeper understanding of themselves, explore new modes of self-expression, and make a positive impact on society. By pursuing their creative aspirations, individuals can tap into their imagination and creativity, and create something unique and meaningful [24].

13.2.5.5 Social aspirations

Social aspirations refer to the goals and ambitions of individuals to make a positive impact on society, to help others, and to improve the quality of life for individuals and communities. Social aspirations are shaped by a variety of factors, such as personal values, life experiences, exposure to different cultures, and awareness of social issues.

Features of social aspirations:

1. Community service: Many individuals aspire to engage in community service and volunteer work, to support social causes and help those in need.
2. Advocacy: Many individuals aspire to become advocates for social issues and use their voices and platforms to raise awareness of important social issues.
3. Leadership: Many individuals aspire to become leaders in their communities, to inspire and mobilize others to create positive change.
4. Sustainability: Many individuals aspire to support sustainability and environmental causes, to reduce their impact on the planet and promote a healthier future for all.
5. Social entrepreneurship: Many individuals aspire to start social enterprises, to use business to create positive social and environmental impact while generating sustainable revenue.

Social aspirations are an important part of social responsibility and civic engagement. By pursuing their social aspirations, individuals can contribute to a better society, build meaningful connections with others, and achieve personal fulfilment resulting in a positive impact on the world around them [25,26].

13.2.5.6 Entrepreneurial aspirations

Entrepreneurial aspirations refer to the desire to start and run a business venture, and these aspirations have become an important component of student aspirations in recent years. In today's rapidly changing economy, where technology is transforming industries and job markets are becoming increasingly competitive, many students are turning to entrepreneurship as a way to create their own career paths and achieve their goals.

Research has shown that the students having higher levels of entrepreneurial aspirations are more likely to become entrepreneurs in the future. A study by Shinnar et al. [27] surveyed over 27,000 students from 20 countries and found that the level of entrepreneurial aspirations among students was a strong predictor of future entrepreneurial activity. Students who had higher levels of entrepreneurial aspirations proved to have started their own businesses and became successful in those ventures.

The development of entrepreneurial aspirations could be influenced by a number of factors, including family background, education, work experience, and exposure to entrepreneurial role models. A study by Liao and Liu [28] found that family background played a significant role in the development of entrepreneurial aspirations among the students. Students from families with an entrepreneurial background were found to have higher levels of entrepreneurial aspirations than those without such a background.

Education also is considered to be an important factor in the development of entrepreneurial aspirations. Entrepreneurship education programs have been found to be effective in increasing entrepreneurial intentions and aspirations among students [29]. Such programs provide students with the knowledge, skills, and resources necessary to start and run a business venture. In addition, exposure to entrepreneurial role models, such as successful entrepreneurs, could also have a positive impact on the development of entrepreneurial aspirations among students.

Furthermore, work experience has been found to be another important factor in the developing the entrepreneurial aspirations. Students who have had work experience in entrepreneurial settings, such as start-ups or small businesses, found to have higher levels of entrepreneurial aspirations [30]. Such work experience provides students with hands-on training in the challenges and rewards of entrepreneurship, and could help them to better equipped with the skills and knowledge required for becoming successful entrepreneurs.

In recent years, entrepreneurial aspirations have become one of the important components of student aspirations. The development of these aspirations can be influenced by many factors, such as family background, education, work experience, and exposure to entrepreneurial role models. Students having high level of entrepreneurial aspirations would likely to become successful entrepreneurs in the future, and the promotion of these aspirations can help to foster a more entrepreneurial culture and drive economic growth [31].

It is important for educators to understand and support student aspirations, as aspirations can be a powerful motivator for learning and achievement. By helping students to identify and pursue their aspirations, educators can be instrumental in making the students develop in their lives a sense of purpose, direction, and fulfilment [16].

13.2.6 Flexibilities

Academic flexibilities refer to the various ways by which educational institutions provide flexibility and choice to the students in their academic pursuits. Academic flexibilities can take many different forms, and they can help students to personalize their educational experiences, meet their unique needs and interests, and achieve their goals [32].

Types of academic flexibilities:

1. Course offerings: Academic institutions may provide a wide range of course offerings, allowing students to choose courses that are of interest to them, and that help them to achieve their academic and career goals.
2. Credit transfer: Academic institutions may allow students to transfer credits from other institutions, or from prior learning experiences, such as work experience, military training, or prior coursework. Academic bank of credits (ABC) is a new concept as mentioned in new education policy (NEP-2020) of India.
3. Flexible scheduling: Academic institutions may offer flexible scheduling options, viz., evening or weekend classes, online courses, and/or self-paced learning programs. This allows students to balance their academic pursuits with other responsibilities and commitments.
4. Dual enrolment: Academic institutions may offer dual enrolment programs by allowing high school students to earn college credits while they are still in high school.
5. Personalized learning plans: Academic institutions may provide personalized learning plans, allowing students to customize their academic paths and focus on areas of interest or need.
6. Multiple exit options: This refers to the flexibility given to learners to exit a program of study at different stages, based on their individual needs and goals. This means that learners can choose to exit the program after completing a certain number of courses, or after earning a specific credential, such as a certificate or diploma.
7. Dual degrees: It refers to the opportunity for learners to pursue two degrees simultaneously, which can be completed in a shorter period of time than pursuing each degree separately. Dual degree programs are designed to provide learners with a broader range of knowledge and skills, as well as greater flexibility and marketability in the job market.
8. Twinning programs: Twinning programmes in academic flexibilities refer to the opportunity for learners to pursue a degree program from two different institutions simultaneously. Twinning programs are designed to provide learners with a broader range of knowledge and skills, as well as exposure to different academic environments, cultures, and perspectives.

9. Semester abroad programs: This flexibility refers to the opportunity for learners to spend a semester studying at a foreign university or educational institution as part of their degree program. Semester abroad programs are basically designed to provide learners an opportunity for an immersive educational and cultural experience, as well as exposure to different academic environments, cultures, and perspectives.

And so, academic flexibilities are important because they allow students to personalize their educational experiences and meet their unique needs and interests. By providing a range of academic flexibilities, the higher educational institutions could help students in achieving their academic and career goals, and to develop the skills and competencies they need to succeed in an ever changing world [33,34].

13.2.7 Industry 5.0

Industry 5.0 is a term used in the recent past to describe the fifth generation of industrial development processes that is characterized by the integration of advanced technologies such as artificial intelligence, machine learning, robotics, and the internet of things (IoT) integrated with human skills and expertise. The goal of Industry 5.0 is to create a sustainable and more human-centered approach to industrial development, where technology is used to enhance and support human skills and capabilities, rather than replacing them.

Industry 5.0 is typically characterized by the following features:

1. Human-machine collaboration: Industry 5.0 emphasizes human-machine collaboration, where in which the advanced technologies such as robotics, artificial intelligence, and machine learning are used to augment human skills and capabilities, rather than replacing them.
2. Sustainable production: Industry 5.0 focuses on sustainable production, where environmental considerations and resource efficiency are prioritized, and waste and emissions are minimized.
3. Customer-centricity: Industry 5.0 emphasizes customer-centricity, where products and services are designed and produced in such a way that they meet the specific needs and preferences of customers, and also provide a personalized and seamless user experience.
4. Smart factories: Industry 5.0 is characterized by the development of smart factories, where advanced and modern technologies such as sensors, data analytics, and automation are used to optimize production processes and improve efficiency.

Industry 5.0 represents a shift towards a more human-centered and sustainable approach to industrial development, where technology is used to enhance and support human skills and capabilities, rather than replacing them. By leveraging advanced technologies and human expertise in a collaborative and sustainable way, Industry 5.0 has the potential to drive innovation, create new business opportunities, and address some of the biggest challenges facing society today [35–40].

13.2.7.1 Human-machine collaboration

Human-machine collaboration refers to the much needed collaboration between humans and machines (e.g., robots, artificial intelligence) for achieving a common goal or a task. In the specific context of industry and work, human-machine collaboration involves leveraging the unique strengths and capabilities of both humans and machines to improve productivity, quality, and safety.

Here are some common examples of human-machine collaboration:

1. Manufacturing: Human-machine collaboration in manufacturing industry involves the use of robots and automation to perform repetitive tasks, while humans are responsible for such tasks, which require flexibility, creativity, and problem-solving.
2. Healthcare: Human-machine collaboration in healthcare involves the use of technology such as robotic surgery, diagnostic imaging, and medical monitoring to assist healthcare professionals in providing high-quality care to patients.
3. Transportation: Human-machine collaboration in transportation involves the use of self-driving cars and trucks, which rely on both human input and artificial intelligence to navigate safely on the road.
4. Education: Human-machine collaboration in education involves the use of technology such as artificial intelligence and virtual reality to create personalized learning experiences for students, while educators provide guidance and support.

Hence, human-machine collaboration is an increasingly important aspect of work and industry, as technology continues to advance and automate many tasks. By leveraging the unique strengths and capabilities of both humans and machines, human-machine collaboration can improve efficiency, safety, and productivity, and create new opportunities for innovation and growth [41,42].

13.2.7.2 Sustainable production

Sustainable production is a manufacturing and production process that prioritizes the efficient use of natural resources, reduces waste and emissions, and promotes environmental sustainability. Sustainable production practices are becoming increasingly important as companies seek to reduce their environmental footprint and meet the demands of consumers who prioritize sustainability.

Here are some common features of sustainable production:

1. Efficient use of resources: Sustainable production emphasizes the efficient and effective use of natural resources such as energy, water, and raw materials. This can be achieved by using efficient technologies, process optimization, and recycling and reuse of materials.
2. Pollution prevention: Sustainable production aims to prevent pollution by reducing waste and emissions, and by using cleaner production technologies and processes.
3. Renewable energy: Sustainable production promotes the use of renewable energy sources, such as solar, wind, and hydro power for reducing greenhouse gas emissions and reliance on non-renewable energy sources.
4. Life cycle thinking: Sustainable production involves considering the entire life cycle of a product (life cycle assessment and analysis), from raw materials to disposal and designing products and production processes that minimize environmental impact at each stage.
5. Stakeholder engagement: Sustainable production involves engaging with stakeholders, including suppliers, customers, and local communities, to promote transparency and collaboration in the production process.

Sustainable production is an important aspect of manufacturing and production that prioritizes environmental sustainability and efficiency. By adopting sustainable production practices, companies can reduce their environmental footprint, improve efficiency, and meet the growing demand for sustainable products and processes [43,44].

13.2.7.3 Customer centricity

Customer centricity is a business philosophy that emphasizes the importance of keeping the customer at the center of all business decisions and activities. In a customer-centric organization, the emphasis is on understanding and meeting the needs and expectations of customers, and providing them with a positive and personalized experience throughout their interactions with the organization.

Here are some common features of a customer-centric approach:

1. Customer understanding: A customer-centric organization understands the needs, wants, and preferences of its customers through research and data analysis.
2. Personalization: A customer-centric organization provides personalized experiences for its customers, such as personalized recommendations, customized products, and tailored marketing messages.

3. Customer feedback: A customer-centric organization solicits feedback from its customers to continuously improve its products, services, and processes.
4. Customer engagement: A customer-centric organization engages with its customers through various channels, such as social media, email, and chatbots, to provide support and build relationships.
5. Continuous improvement: A customer-centric organization continuously improves its products, services and processes in a way that they meet the exact changing needs and actual expectations of its customers.

Customer centricity is an important aspect of business strategy that emphasizes the importance of meeting the needs and expectations of customers. By adopting a customer-centric approach, organizations can improve customer loyalty, increase revenue, and gain an edge over the others competitors in the marketplace [45–47].

13.2.7.4 Smart factories

Smart factories are advanced manufacturing facilities that leverage digital technologies [Internet of Things (IoT), artificial intelligence (AI), and automation] to optimize production processes and to improve the operational efficiency. Smart factories are designed to be highly flexible and adaptable and can respond quickly to ever changing market demands and customer needs.

Here are some common features of smart factories:

1. Connected machines and devices: Smart factories use connected machines and devices to gather and share data in real-time, enabling greater visibility and control over the production process.
2. Automation and robotics: Smart factories use automation and robotics to perform repetitive tasks and improve productivity, while reducing the physical load on human workers so as to enable them divert their energies in solving more complex problems and addressing creative tasks.
3. Predictive maintenance: Smart factories use predictive maintenance techniques to detect potential issues with equipment and machinery before they occur, reducing downtime and increasing productivity.
4. Digital twin technology: Smart factories use digital twin technology to create virtual models of the production process, enabling testing and optimization of new products and processes in a virtual environment.
5. Data analytics and AI: Smart factories use data analytics and artificial intelligence (AI) to optimize production processes, to improve quality control and to identify areas for improvement.

Smart factories are designed to be highly efficient and adaptable, and can help manufacturers to achieve higher levels of productivity, reduce costs, and improve product quality. By leveraging the power of digital technologies, smart factories are in the process of transforming the entire manufacturing industry and also paving the way for a more connected and efficient future [48,49].

13.2.8 Curriculum design

Curriculum design is the process of creating a comprehensive plan for teaching and learning that outlines the content, skills, and competencies that students are expected to learn. Curriculum design involves a systematic approach to designing, implementing, and evaluating educational programs that are aligned with specific learning objectives and goals.

Here are the key steps involved in curriculum design:

1. Needs assessment: To identify the needs and goals of the learners, the community, and the educational system.
2. Learning objectives: To develop clear and measurable learning objectives that specify what learners should be able to do or know by the end of the educational program.
3. Content selection: To select and organize the content that will be taught in the program, based on the learning objectives and the needs of the learners.
4. Teaching strategies: Choose teaching strategies that are appropriate for the content and learning objectives, and that engage learners in active and meaningful learning.
5. Assessment: To develop appropriate methods for assessing learners' progress and achievement and to use the results to evaluate the effectiveness of the curriculum.
6. Revision: Continuously evaluate and revise the curriculum to ensure that it remains relevant and effective.
7. Benchmarking: 70% of the curriculum design (framework, structure, courses, contents, etc.) are to be incorporated by taking from the best of the benchmarked institutions globally and 30% of it should be unique to the parent institution.
8. Digital lean solutions: Digital lean solutions are an innovative approach to improving business operations by combining digital technology (SMAC), information technology, and Internet of Things (IoT).

Curriculum design is a complex and iterative process that requires careful planning, implementation, and evaluation. By following a systematic approach to curriculum design, educators can develop effective educational programs, which meet the needs and goals of learners and prepare them for success in their personal and professional lives [50].

13.2.8.1 Assessment of needs

Assessment of needs is an important part of curriculum design, as it helps to ensure that the curriculum is tailored to the needs of target audience, such as students or learners. Needs assessment in curriculum design involves identifying the learning needs and goals of the target audience, as well as the resources and constraints that may impact the design and the implementation of the curriculum.

Steps involved in needs assessment as a part of curriculum design:

1. Identify the target audience: The first step in needs assessment in curriculum design is to identify the target audience, such as students, professionals, or adult learners.
2. Determine the purpose: The next step is to determine the purpose of the curriculum, such as to provide basic education, to develop specific skills, or to prepare for a particular profession.
3. Collect data: The third step is to collect the data using various methods including surveys & interviews with stake holders, discussion with the focus groups, and observation to identify the learning needs and goals of the target audience.
4. Data Analysis: The next step is analyzing the data collected to determine the gaps between the current knowledge and skills of the target audience and the desired learning outcomes.
5. Develop the curriculum: Based on the results obtained from the needs assessment, a curriculum is developed that addresses the identified gaps in knowledge and skills. The curriculum should be designed to be engaging, relevant, and appropriate for the target audience.
6. Implement and evaluate: Finally, the curriculum is implemented and evaluated to determine its effectiveness in achieving the desired learning outcomes. Based on the evaluation results, adjustments may be made to the curriculum to improve its effectiveness.

Needs assessment is a critical process in curriculum design, as it helps to ensure that the curriculum is mapped to the needs of the target audience and is very effective in achieving the desired and designed learning outcomes. By using data-driven approaches to identify the learning needs and goals of the target audience, curriculum designers can create engaging and effective curricula that prepare learners for success in their chosen field [51,52].

13.2.8.2 Learning

Learning is a key component of curriculum design, as it forms the basis for the educational experiences and outcomes that the curriculum is intended to achieve. Learning in curriculum design refers to the process of acquiring new knowledge, skills and understanding through various educational activities and experiences.

Some key factors to consider when designing learning experiences in curriculum design:

1. Learning objectives: Learning objectives are specific goals that are established for a particular course or program of study. These objectives should be aligned with the overall goals and the outcomes of curriculum.
2. Teaching methods: Teaching methods refer to the techniques and strategies that are utilized to deliver the content of the curriculum and to facilitate the learning process. Effective teaching methods include lectures, discussions, hands-on activities, and collaborative learning.
3. Assessment methods: Assessment methods are used to evaluate the learning outcomes of curriculum. Assessment methods should be aligned with the learning objectives and should include a variety of methods such as exams, essays, projects, and presentations.
4. Learning resources: Learning resources are the tools and materials that are used to support learning. These resources may include textbooks, online materials, multimedia resources, and learning management systems.
5. Learning environment: The learning environment refers to the physical and social setting in which learning takes place. This may include the classroom, laboratory, online environment, or community settings.
6. Learner-centered approach: Curriculum design should focus on a learner-centered approach, which places the needs and interests of learner at the center of the learning experience. This includes providing opportunities for self-directed learning, individualized learning, and collaborative learning.

Effective learning is essential to the success of any curriculum design. By focusing on the needs and interests of learner and designing engaging and effective learning experiences, curriculum designers can help learners achieve their educational goals and prepare them for success in their chosen field [53,54].

13.2.8.3 Content selection

Content selection is a crucial part of curriculum design, as it involves selecting the most relevant and appropriate content to achieve the desired learning outcomes. Content selection should be based on the needs and interests of target audience, as well as the goals and outcomes of the curriculum.

Some key factors to consider when selecting content for curriculum design:

1. Learning objectives: The content selected should be aligned with the learning objectives of curriculum. This ensures that the content is relevant and appropriate for the desired learning outcomes.
2. Relevance: The content selected should be relevant to the needs and interests of target audience. This ensures that the content is engaging and meaningful for learners.
3. Quality: The content selected should be of high quality and should be accurate, up-to-date, and reliable.
4. Diversity: The content selected should represent diverse perspectives and experiences for providing learners a well-rounded education.
5. Balance: The content selected should be balanced, covering a range of topics and perspectives without bias.
6. Appropriateness: The content selected should be appropriate for age, developmental level, and cultural background of the target audience.
7. Accessibility: The content selected should be accessible to all learners regardless of their background and/or abilities.

Content selection is a critical part of curriculum design, as it can greatly impact the effectiveness and relevance of the curriculum. By selecting high-quality, relevant, and diverse content that is aligned with the learning objectives of the curriculum, curriculum designers can create engaging and effective learning experiences that prepare learners for success in their chosen field [55–57].

13.2.8.4 Teaching strategies

Teaching strategies are an important component of curriculum design, as they are used to deliver the curriculum content and facilitate learning. Effective teaching strategies should be aligned with the specified learning objectives and needs & interests of the target audience.

Some common teaching strategies used in curriculum design:

1. Lectures: Lectures are a traditional teaching strategy that involves presenting information through spoken word. Lectures can be effective for introducing new concepts and providing an overview of a topic.
2. Discussions: Discussions are an interactive teaching strategy that allows learners to engage in conversation and share their ideas and perspectives. Discussions can be effective for promoting critical thinking and collaboration.
3. Hands-on activities: Hands-on activities involve learners in active participation and can be effective for promoting experiential learning and skill development.

4. Collaborative learning: Collaborative learning involves learners working together in small groups to solve problems, complete projects, or discuss ideas. Collaborative learning can be effective for promoting teamwork and communication skills.
5. Inquiry-based learning: Inquiry-based learning involves learners posing questions, investigating topics, and seeking answers through research and exploration. Inquiry-based learning can be effective for promoting curiosity and self-directed learning.
6. Technology-based learning: Technology-based learning involves using digital tools and resources for enhancing the teaching and learning processes. Technology-based learning can be effective for providing access to information and promoting multimedia learning experiences.

Effective teaching strategies are essential to the success of curriculum design. By selecting teaching strategies that are aligned with the learning objectives and needs & interests of the target audience, curriculum designers can create engaging and effective learning experiences that prepare learners for success in their chosen field [58–60].

13.2.8.5 Assessment

Assessment is a critical component of curriculum design, as it is used to evaluate the effectiveness of the curriculum and learning outcomes achieved by the learners. Assessment should be aligned with the learning objectives and teaching strategies used in the curriculum.

Common types of assessment used in curriculum design:

1. Formative assessment: Formative assessment is used to monitor learner progress and provide feedback during learning process. Formative assessment can be used to identify areas where learners need additional support or to adjust the teaching strategies to meet the needs of learners.
2. Summative assessment: Summative assessment is used to evaluate the learning outcomes achieved by learners at the end of a course or a program of study. Summative assessment could be in the form of examinations, essays, projects, or presentations.
3. Authentic assessment: Authentic assessment involves evaluating the ability of learners in applying knowledge and skills in real-world situations. Authentic assessment can be used to evaluate learners' problem-solving skills, critical thinking skills, and creativity.
4. Self-assessment: Self-assessment involves learners evaluating their own learning progress and achievements. Self-assessment can be used to encourage learners to take ownership of their learning and to develop self-reflection skills.

5. Peer assessment: Peer assessment involves learners themselves evaluating the work of their peers. Peer assessment can be used to promote collaboration, teamwork, and communication skills.
6. Portfolio assessment: Portfolio assessment involves learners compiling their work over time to demonstrate the progress in their learning and achievements. Portfolio assessment can be used to provide a comprehensive picture of learners' knowledge and skills.

Effective assessment is essential to the success of curriculum design. By using a variety of assessment methods that are aligned with the learning objectives and teaching strategies, curriculum designers can evaluate the effectiveness of curriculum and learning outcomes achieved by the learners, and make adjustments as needed to improve curriculum and enhance learning experience [61–65].

13.2.8.6 Revision

Revision is an important part of curriculum design, as it allows curriculum designers to evaluate the effectiveness of curriculum and make necessary adjustments to better meet the needs of learners and to achieve desired learning outcomes.

Key steps involved in revision in curriculum design:

1. Evaluate the effectiveness: The first step in revision is to evaluate the effectiveness of the curriculum by reviewing assessment data, feedback from learners, and other performance indicators.
2. Identify areas for improvement: Based on the evaluation, identify areas where the curriculum can be improved, such as updating content, revising learning objectives, or adjusting teaching strategies.
3. Develop a plan for revision: Develop a plan for revision that outlines the changes to be made, the resources needed, and the timeline for implementation.
4. Implement the revision: Implement the revision by updating the curriculum content, adjusting teaching strategies, and providing necessary resources and support for learners.
5. Evaluate the impact: Evaluate the impact of the revision by reviewing assessment data, feedback from learners, and other performance indicators. This will help to determine the effectiveness of the revision and identify any further areas for improvement.

Revision is an ongoing process in curriculum design that is essential for ensuring that the curriculum is effective in achieving the desired learning outcomes. By regularly evaluating the effectiveness of the curriculum and making necessary adjustments, curriculum designers can create engaging and effective learning experiences that prepare learners for success in their chosen field [55,66].

13.2.8.7 Benchmarking in curriculum design

Benchmarking in curriculum design involves comparing an educational program to other similar programs to identify areas for improvement and ensure that the program meets the required standards and expectations. Benchmarking can be used to compare an educational program to other programs at the same level (e.g. high school to high school) or to programs at higher levels (e.g. high school to college).

Here are some common steps involved in benchmarking in curriculum design:

1. Identify the benchmark: Identify the program or programs that will be used as the benchmark for comparison. These programs should be similar in terms of content, goals, and outcomes to the educational program being benchmarked.
2. Collect data: Collect data on the benchmark program(s), such as curriculum documents, syllabi, course descriptions, and student learning outcomes.
3. Analyze data: Analyze the data collected from the benchmark program(s) for identifying strengths, weaknesses, and areas for improvement.
4. Compare to own program: Compare the data collected from the benchmark program(s) to the educational program being benchmarked to identify similarities and differences.
5. Adjust the curriculum: Use the information gathered from the benchmarking process to adjust and improve the educational program as needed.

Benchmarking in curriculum design can help ensure that an educational program is meeting the necessary standards and expectations, and can identify areas for improvement. By comparing an educational program to similar programs, educators can gain insights into effective teaching and learning strategies and make data-driven decisions to improve the program [67–70].

13.2.8.8 Digital lean solutions

Digital lean solutions are an innovative approach to improving business operations by combining digital technology (SMAC), information technology, and Internet of Things (IoT). This approach is basically transforming the way businesses operate by streamlining processes, increasing efficiency, and reducing waste. These digital lean solutions along the lines of lean manufacturing systems are going to change the way we live in the very near future. Hence, exposure to these solutions need to be given to the students of all disciplines irrespective of their domain in higher education sector. Let us explore this concept further and examine its potential benefits.

13.2.8.8.1 Core components of digital lean solutions

Digital Technology (SMAC): SMAC stands for Social networks, Mobile network, Analytics, and Cloud. These technologies are essential in the digital age and enable businesses to streamline processes, improve customer experiences, and make better decisions.

Information Technology: IT is the backbone of any digital lean solution. It involves the use of computers, software, and other technologies to manage and process information.

Internet of Things (IoT): IoT involves the interconnectivity of physical devices and systems through the internet. This enables real-time data collection and analysis that can be used to optimize the business operations.

13.2.8.8.2 Benefits of digital lean solutions

Improved Efficiency: By streamlining processes and reducing waste, digital lean solutions can help businesses become more efficient and effective. This can lead to cost savings and increased productivity.

Enhanced Customer Experiences: With the use of digital technology, businesses can better understand customer needs and preferences, which can lead to personalized experiences and improved customer satisfaction.

Real-Time Data Analysis: The IoT enables businesses to collect the real-time data that can be used to optimize processes and to make better decisions.

Greater Flexibility: Digital lean solutions enable businesses to be more agile and responsive to changing market conditions, which can help them stay competitive.

Examples of digital lean solutions in action include:

Smart Manufacturing: By combining IoT sensors, data analytics, and automation, manufacturers can optimize their operations and reduce waste.

Connected Logistics: By using real-time data and analytics, logistics companies can optimize their supply chain and reduce costs.

Smart Retail: By using digital technology, retailers can provide personalized experiences to customers, improve inventory management, and increase sales.

In conclusion, digital lean solutions are a powerful tool for businesses to streamline operations, improve customer experiences and to stay competitive in the digital age. By combining digital technology (SMAC), information technology, and the IoT, businesses can optimize their operations and make better decisions. As such, organizations that adopt digital lean solutions are better positioned to thrive in the future [71–73].

13.2.9 Teaching-learning-evaluation processes

Teaching, learning, and evaluation are three interrelated components of the educational process. Effective teaching involves creating learning experiences that engage and motivate students, while effective learning involves actively participating in these experiences to gain knowledge and develop skills. Evaluation helps to determine whether the learning objectives are achieved or not and also provides feedback to improve future teaching and learning [64].

Here is how teaching, learning, and evaluation are interrelated:

1. Teaching: Effective teaching involves creating a supportive learning environment, setting clear learning objectives, selecting appropriate instructional strategies, and providing feedback and support to learners.
2. Learning: Effective learning involves actively engaging in the learning experience, developing critical thinking and problem-solving skills, and applying new knowledge and skills to real-world situations.
3. Evaluation: Evaluation helps to determine whether the learning objectives have been achieved and also provides feedback to improve future teaching and learning processes. Evaluation can take many forms such as tests, projects, presentations, or portfolios and can involve both formative (ongoing) and summative (final) assessments.

13.2.9.1 Teaching

Teaching is a critical part of teaching-learning-evaluation process and involves the facilitation of learning experiences for students. Teaching is a complex process that involves a wide range of skills and strategies including lesson planning, classroom management, delivery of instructions, and assessment of student learning.

Effective teaching is characterized by a student-centric approach in which the teacher works collaboratively with students to create a positive and engaging learning environment. This approach emphasizes the importance of understanding the individual needs and learning styles of each student, and adapting instruction and learning activities accordingly. It also involves the use of a variety of teaching strategies and techniques such as experiential learning, project-based learning, and technology-enhanced learning.

One of the key aspects of effective teaching is the ability to design and deliver engaging and meaningful lessons that are aligned with the learning objectives of curriculum. This involves a systematic approach to lesson planning, which includes the identification of learning outcomes, selection of appropriate teaching strategies and materials, and assessment of student learning.

Another important aspect of effective teaching is classroom management. This involves creating a safe and supportive learning environment that is conducive to the process of learning. Classroom management strategies include establishing clear expectations and rules, providing positive reinforcement, and using appropriate disciplinary measures when necessary.

The delivery of instruction is also a critical part of effective teaching. This involves the use of a variety of instructional methods and techniques to engage students and promote learning. These may include lectures, discussions, group activities, and hands-on experiences.

Finally, the assessment of student learning is an essential part of effective teaching. Assessment is used to evaluate student understanding and progress and to provide feedback for improvement. Assessment strategies may include formative assessments, such as quizzes and homework assignments, as well as summative assessments, such as exams and final projects.

Effective teaching is a critical component of the teaching-learning-evaluation process. It involves the use of a wide range of skills and strategies including lesson planning, classroom management, delivery of instruction, and assessment of student learning. Effective teaching requires a student-centered approach that emphasizes the importance of understanding individual student needs and learning styles, and adapting instruction and learning activities accordingly. With effective teaching, students can be engaged and motivated to learn and can achieve their full potential [74–77].

13.2.9.2 Learning

Learning is another important part of the teaching-learning-evaluation process, and it involves the acquisition of knowledge, skills, and attitudes by students. Learning is a complex process, which is influenced by a wide range of factors including the quality of teaching, the learning environment, the student's motivation and engagement, and the use of appropriate learning strategies.

Effective learning is characterized by a student-centric approach in which the student takes an active role in the learning process. This approach emphasizes the importance of understanding individual learning needs and styles and adapting learning activities and strategies accordingly. Effective learning also involves the development of critical thinking skills, problem-solving skills, and the ability to apply knowledge to real-world situations.

One of the key aspects of effective learning is development of metacognitive skills that enable students to reflect on their own learning and identify areas for improvement. This involves setting goals, monitoring progress, and evaluating the effectiveness of learning strategies.

Another important aspect of effective learning is the use of active learning strategies such as group work, discussions, and project-based learning. Active learning strategies promote engagement, collaboration, and the development of higher-order thinking skills.

The learning environment also is an important factor in effective learning. A positive and supportive learning environment could promote motivation and engagement while a negative or hostile environment can have the opposite effect. A positive learning environment includes features such as clear expectations, opportunities for collaboration and interaction, and a variety of resources and learning materials.

Finally, feedback is an important part of effective learning. Feedback can provide students with information on their progress and performance and help them to identify areas for improvement. Effective feedback is timely, specific, and constructive, and it should be delivered in a supportive and non-judgmental manner.

Effective learning is one of the important components of teaching-learning-evaluation process. It involves a student-centered approach that emphasizes the importance of understanding individual learning needs and styles, and adapting learning activities and strategies accordingly. Effective learning also involves the development of metacognitive skills, the use of active learning strategies, the creation of a positive learning environment, and the provision of constructive feedback. With effective learning, students can acquire the knowledge, skills, and attitudes necessary to succeed in today's rapidly changing world [78–80].

13.2.9.3 Evaluation

Evaluation is the most critical part of the teaching-learning-evaluation process and it involves the assessment of student learning and the effectiveness of teaching strategies. Evaluation is a complex process that can be influenced by a wide range of factors including the quality of teaching, the learning environment, and the student's motivation and engagement.

Effective evaluation is characterized by a student-centered approach in which the focus is on the student's learning and progress rather than just their grades and/or test scores. This approach emphasizes the importance of assessing not only what the student has learned, but also how they learned and the skills and attitudes they developed.

One of the key aspects of effective evaluation is the use of multiple assessment strategies including formative and summative assessments. Formative assessments are used throughout the learning process to provide feedback to the student and the teacher on their progress and to guide future learning activities. Summative assessments are used at the end of a learning period to evaluate the student's overall achievement and understanding.

Another important aspect of effective evaluation is the use of authentic assessments, which measure the student's ability to apply knowledge and skills to real-world situations. Authentic assessments can include projects, presentations, and other performance-based tasks that require students to demonstrate their understanding and application of concepts and skills.

The evaluation process also involves the use of rubrics and criteria to ensure that assessments are fair, consistent, and aligned with learning objectives. Rubrics provide a clear and transparent framework for assessment, and criteria ensure that assessments are aligned with the specific learning objectives and standards.

Finally, effective evaluation involves the use of feedback to inform future learning and teaching strategies. Feedback can be provided to the student on their progress and performance, and to the teacher on the effectiveness of their teaching strategies. Feedback can also be used to guide future learning activities and to identify areas for improvement.

Effective evaluation is an important component of the teaching-learning-evaluation process. It involves a student-centered approach that emphasizes the importance of assessing not only what the student has learned, but also how they learned and the skills and attitudes they developed. Effective evaluation involves the use of multiple assessment strategies, including authentic assessments and rubrics, and the provision of feedback to inform future learning and teaching strategies. With effective evaluation, students can be empowered to achieve their full potential and succeed in a rapidly changing world [81–84].

To summarize, Teaching, learning, and evaluation are interdependent and must be carefully planned and implemented to ensure the success of educational process. By using effective teaching strategies, promoting active learning, and providing meaningful evaluation, educators can create positive learning environment that engages students and helps them achieve their full potential.

13.2.10 Outcome-based education (OBE)

Outcome-based education is an approach to education, which focuses on identifying and achieving specific learning outcomes or goals. In outcome-based education, the curriculum is designed to ensure that students acquire the knowledge, skills, and competencies they need to succeed in their personal and professional lives.

The 'Washington Accord is an international accreditation agreement for professional engineering academic degrees between the bodies responsible for accreditation in its signatory countries and regions'. Established in 1989, the full signatories as of 2018 are Australia (1989), Canada (1989), China, Hong Kong, India (2014), Ireland, Japan, Korea, Malaysia, New Zealand (1989), Pakistan, Peru, Philippines, Russia, Singapore, South Africa, Sri Lanka, Taiwan, Turkey, the United Kingdom (1989) and the United States (1989). However, Bangladesh, Chile, Costa Rica, Mexico, and Philippines do have provisional signatory status and might become member signatories in the near future.

In line with the Washington Accord, the graduate attributes have been specified for engineering education which are applicable with sight modifications for all the programs of study in University education systems across the globe.

In Outcome-based education (OBE) system, starting from the Vision & Mission Statements of University/Institute, School, and Department, the Graduate Attributes (As per Washington Accord) (GAs), Program Educational Objectives (PEOs), Program Outcomes (POs) & Program Specific Outcomes (PSOs) and Course Outcomes (COs) should be in sync and mapped to every program of study. It is not enough to specify the above outcomes, but it is equally important that the outcomes are measured through Student evaluation components (Direct measurement) & Surveys (Indirect measurement) of all the stake holders (direct stake holders such as students and alumni, and indirect stake holders such as parents, employers, and prospective students).

Here are some common features of outcome-based education:

1. Learning outcomes: Outcome-based education emphasizes the importance of identifying clear and measurable learning outcomes that specify what students should be able to do or know by the end of an educational program.
2. Competency-based approach: Outcome-based education uses a competency-based approach, where students are expected to demonstrate mastery of specific skills and competencies.
3. Assessments: Outcome-based education involves ongoing assessment, where students are evaluated based on their ability to demonstrate the desired learning outcomes and competencies.
4. Flexibility: Outcome-based education allows for flexibility in learning process as students are encouraged to take ownership of their learning and pursue their own interests and goals.
5. Collaboration: Outcome-based education emphasizes collaboration between educators and students as well as between students themselves to achieve the desired learning outcomes and competencies.

Overall, outcome-based education is an approach to teaching and learning processes that emphasizes the importance of clear and measurable learning outcomes and the development of specific skills and competencies. By using appropriate assessments to evaluate student progress and by adjusting teaching strategies as needed, outcome-based education can help ensure that students are well-prepared to succeed in their personal and professional lives [85].

13.2.11 Teacher-centric learning vs. student-centric learning

Teacher-centric learning and student-centric learning are two different approaches to education that have different goals and methods.

Teacher-centric learning, also known as traditional or lecture-based learning, is a method of education where the teacher is the central figure in the classroom. The teacher is responsible for delivering the content and the students are expected to listen, take notes, and memorize the material. The teacher controls the pace and direction of the learning and students are assessed based on their ability to reproduce what they have learned.

Student-centric learning, also known as learner-centered or active learning, is a method of education where the focus is on the needs and interests of the student. In this approach, the teacher acts as a facilitator rather than a person delivering lectures. Students are encouraged to take active role in their learning by participating in discussions, collaborative activities and projects. Assessment is focused on the students' ability to apply what they learned to real-world situations.

Key differences between teacher-centric learning and student-centric learning:

1. Role of the teacher: In teacher-centric learning, the teacher is the central figure in the classroom, responsible for delivering the content and controlling the learning process. In student-centric learning, the teacher acts as a facilitator rather than a person professes, guiding, and supporting the students in their learning.
2. Learning outcomes: In teacher-centric learning, the focus is on acquiring knowledge and reproducing it. In student-centric learning, the focus is on developing skills and competencies that can be applied in real-world contexts.
3. Assessment: In teacher-centric learning, assessment is focused on the students' ability to reproduce what they have learned. In student-centric learning, assessment is focused on the students' ability to apply what they learned to real-world contexts.
4. Learning environment: In teacher-centric learning, the learning environment is often a traditional classroom setting, where teacher lectures and students listen. In student-centric learning, the learning environment is often more interactive and collaborative, with students working in groups and engaging in discussions and projects.

Both teacher-centric learning and student-centric learning have their advantages and disadvantages. However, student-centric learning is becoming increasingly popular as it is seen as a more effective way to prepare students for achieving success in today's rapidly changing world scenarios [86].

13.3 EDUCATION 5.0

13.3.1 Background

Education 5.0 encompasses all the above aspects and seamlessly integrates all of them.

Education 5.0 is a relatively new concept that refers to the fifth generation of educational models, which aims to address the changing needs of modern society. It is based on the idea of lifelong learning, where education is not limited to the traditional classroom setting but can happen anytime and anywhere [1].

The first four generations of educational models are:

1. Education 1.0: Education 1.0 refers to the traditional teacher-centric model of education, where the teacher is the primary source of knowledge and learners are passive recipients of information.
2. Education 2.0: Education 2.0 is a more student-centered model of education, where learners are encouraged to take active role in their learning. This can include collaborative learning activities, peer-to-peer mentoring, and project-based learning.
3. Education 3.0: Education 3.0 is a model of education that emphasizes the use of 'digital technologies' to support learning. This can include online courses, educational apps, and virtual learning environments.
4. Education 4.0: Education 4.0 is a model of education that focuses on the development of 21st-century skills such as critical thinking, creativity, collaboration, and communication. Education 4.0 also emphasizes the importance of lifelong learning and the need for learners to adapt to the ever changing world.

These four models of education represent a progression from a traditional teacher-centered approach to a more student-centered, technology-enabled, and future-oriented approach to education.

13.3.1.1 Education 1.0

Education 1.0 refers to the traditional model of education, which is characterized by a teacher-centered approach to learning. In Education 1.0, teacher is the primary source of knowledge and learners are passive recipients of information. The focus is on transmitting knowledge from the teacher to the learner, rather than on active engagement and participation.

Education 1.0 is typically characterized by the following features:

1. Lecture-based teaching: Education 1.0 relies on lecture-based teaching, where the teacher presents information to the class through lectures, and learners take notes and memorize the information.
2. Passive learning: Education 1.0 is characterized by passive learning, where learners are expected to listen, memorize, and regurgitate information, without much opportunity for critical thinking or active engagement.
3. Standardized testing: Education 1.0 relies heavily on standardized testing, which measures learners' ability to memorize and recall information.

4. Classroom-based learning: Education 1.0 is typically delivered in a classroom setting, where learners attend classes at set times and follow a structured curriculum.

Education 1.0 is a traditional and teacher-centered approach to education, which emphasizes the transmission of knowledge from teacher to learner rather than on active engagement and participation [87].

13.3.1.2 Education 2.0

Education 2.0 is a model of education that emphasizes a more student-centered approach to learning, where learners are encouraged to take active role during their learning process. In Education 2.0, learners are viewed as active participants during the learning process and the focus is on creating a collaborative and interactive learning environment.

Education 2.0 is typically characterized by the following features:

1. Active learning: Education 2.0 emphasizes active learning, where learners take active role during their learning process through activities such as collaborative learning, peer-to-peer mentoring, and project-based learning.
2. Personalized learning: Education 2.0 is designed to be more personalized, where learners have more control over their learning experience and can choose the learning path that best suits their needs and interests.
3. Technology-enabled learning: Education 2.0 leverages technology to support learning, such as online learning platforms, educational apps, and virtual learning environments.
4. Collaborative learning: Education 2.0 promotes collaborative learning, where learners work together as a group to solve problems, share ideas, and learn from each other.

Education 2.0 is a more student-centered, interactive, and collaborative approach to learning, which emphasizes the importance of active engagement and participation, personalized learning, and the use of modern technology to support learning [88].

13.3.1.3 Education 3.0

Education 3.0 is a model of education that focuses on integration of technology into learning process and use of digital tools to support and enhance learning. In Education 3.0, learners are encouraged to use technology to access information, communicate with others, and create and share knowledge.

Education 3.0 is typically characterized by the following features:

1. Online learning: Education 3.0 emphasizes the use of online learning platforms, such as 'learning management systems (LMS)' and 'massive open online courses (MOOCs)', to deliver courses and educational content to learners around the world.
2. Flipped learning: Education 3.0 promotes flipped learning, where learners watch instructional videos and complete online activities outside of class and use class time for collaborative and interactive learning activities.
3. Digital literacy: Education 3.0 emphasizes the importance of digital literacy, where learners are taught to use digital tools and platforms effectively and responsibly.
4. Open education: Education 3.0 promotes open education, where educational resources are freely available to learners around the world and learners are encouraged to contribute to creation and sharing of knowledge.

Education 3.0 is a model of education that emphasizes integration of technology into learning process and the use of digital tools to support and enhance learning. It is designed to provide learners with access to a wealth of educational resources and opportunities and to prepare them for the digital age [89,90].

13.3.1.4 Education 4.0

Education 4.0 is a model of education that focuses on developing 21st-century skills such as critical thinking, creativity, collaboration, and communication. In Education 4.0, learners are encouraged to be active learners, and to develop the skills and competencies they need to succeed in today's ever changing world.

Education 4.0 is typically characterized by the following features:

1. Personalized learning: Education 4.0 emphasizes personalized learning, where learners have more control over their learning experience and can choose the learning path that best suits their needs and interests.
2. Competency-based learning: Education 4.0 focuses on competency-based learning, where learners are assessed based on their ability to demonstrate their mastery of specific skills and competencies.
3. Lifelong learning: Education 4.0 promotes lifelong learning, where learners are encouraged to continue learning throughout their lives, and to adapt to a rapidly changing world.
4. Real-world learning: Education 4.0 emphasizes real-world learning, where learners are provided with opportunities to apply their knowledge and skills to real-world problems and situations successfully.

Education 4.0 is a model of education that is designed to develop skills and competencies learners need to succeed in the 21st century. It emphasizes the importance of active learning, personalized learning, competency-based learning, and real-world learning. It also encourages learners to be lifelong learners who are prepared to adapt to a rapidly changing world [1,91,92].

13.3.2 Education 5.0

Education 5.0 is the latest evolution of educational models, which emphasizes the use of technology, personalization, collaboration, and experiential learning to prepare students for success in today's ever changing world. Education 5.0 builds on the previous generations of educational models, which focused on the transmission of knowledge and skills, to create a more holistic and student-centric approach to learning.

Education 5.0 is a model of education that builds on the previous models of education (Education 1.0, 2.0, 3.0, and 4.0) and integrates them with new developments and emerging technologies to create a more holistic, student-centric, and future-oriented approach to education.

Some of the key features of Education 5.0 include:

1. Personalization: Education 5.0 is designed to be personalized to meet the needs of individual learner. This means that learners can choose the type of learning experience that best suits their individual needs and preferences.
2. Collaboration: Education 5.0 emphasizes collaboration and teamwork, where learners work together to solve problems and achieve learning goals.
3. Technology: Education 5.0 incorporates the latest technology to support learning, including online platforms, virtual reality, and artificial intelligence.
4. Experiential learning: Education 5.0 emphasizes hands-on experiential learning, where learners can apply what they learned to real-world situations.
5. Globalization: Education 5.0 emphasizes the importance of global learning, where learners can gain an understanding of different cultures and perspectives.

Education 5.0 is designed to be more flexible, personalized and collaborative than traditional educational models. It aims to prepare learners for the challenges of the 21st century and beyond by providing them with the skills and knowledge they need to succeed in this ever changing world [1,93–97].

13.3.2.1 Personalization

Personalization is a key feature of Education 5.0, which recognizes that every student has unique learning needs and preferences. Personalized learning experience can be created using technologies such as artificial intelligence and adaptive learning algorithms that allow educators to tailor the content, pace, and style of instruction to individual students.

Personalization aims to provide learners with a more tailored and individualized learning experience. In Education 5.0, personalization means that learning is customized to meet the unique needs and interests of each learner based on their individual learning style, preferences, strengths, and weaknesses.

There are many ways in which personalization can be implemented in Education 5.0. For example:

1. Customized learning paths: Learner can choose his/her own learning path based on his/her interests, goals, and level of proficiency. This can include selecting courses, modules or projects that are aligned with their career aspirations or personal interests.
2. Adaptive learning: Education 5.0 leverages technology to provide learners with personalized learning experiences. This can include adaptive learning systems that adjust the difficulty, content, and pace of learning based on the learner's performance and progress.
3. Flexible scheduling: Education 5.0 allows learners to learn at their own pace and on their own schedule using a variety of formats and modalities such as online courses, virtual classrooms, and/or blended learning environments.
4. Personalized assessments: Education 5.0 uses a range of assessments to measure learner progress and provide personalized feedback. This can include self-assessments, peer assessments, and instructor feedback.

13.3.2.1.1 Customized learning paths

'Customized learning paths' is a critical component of the personalization approach to education that is a key feature of Education 5.0. Customized learning paths allow students to tailor their learning experience to their individual needs, interests, and goals. This approach recognizes that every student is unique and has their own learning style, pace, and preferences.

Customized learning paths involve the use of adaptive learning technologies that allow students to work in their own way and at their own pace. These technologies use algorithms and analytics to assess a student's progress and to adjust learning activities and resources accordingly. This ensures that students are receiving instruction that is appropriate to their level of understanding, and that challenges them appropriately.

One of the key benefits of customized learning paths is that they enable students to take ownership of their learning process. Students are empowered to make choices about what and how they learn, which can help to increase engagement and motivation. Customized learning paths also allow for greater flexibility in scheduling and pacing, which can help students to balance their academic work with other commitments and activities.

Another benefit of customized learning paths is that they can help to improve student outcomes. By providing students with targeted instruction and support, customized learning paths can help to close achievement gaps and improve student performance. Customized learning paths can also help to develop critical thinking skills, problem-solving skills, and the ability to apply knowledge to real-world situations.

However, there are also some challenges associated with customized learning paths. One of the biggest challenges is ensuring that students receive adequate support and guidance. While customized learning paths can be highly engaging and motivating, students may need additional support from teachers and mentors to ensure that they are on track and receiving the support they need.

Another challenge is the need for effective assessment and evaluation. Customized learning paths can make assessment more challenging, as students may be working on different tasks and at different levels. It is important to ensure that assessments are aligned with learning objectives and standards and that they are fair and consistent across different students and learning paths.

Customized learning paths is an important component of the personalization approach to education that is a key feature of Education 5.0. Customized learning paths enable students to tailor learning experience to their individual needs, interests, and goals. While there are some challenges associated with customized learning paths, when implemented effectively, they can help to improve student outcomes and prepare students for success in this fast changing world [98–100].

13.3.2.1.2 Adaptive learning

Adaptive learning is another important component of the personalization approach to education that is a key feature of Education 5.0. Adaptive learning refers to the use of technology and data analytics to personalize learning experiences for individual students. This approach recognizes that every student is unique and has their own learning style, pace, and preferences.

Adaptive learning involves the use of algorithms and analytics to assess a student's progress and to adjust learning activities and resources accordingly. This ensures that students are receiving instruction that is appropriate to their level of understanding, and that challenges them appropriately. Adaptive learning technologies can also provide real-time feedback to students, allowing them to monitor their progress and adjust their learning strategies as needed.

One of the key benefits of adaptive learning is that it enables a student to work at his/her own pace and in his/her own way. Students can progress through learning activities and assessments at a pace that is appropriate for them, and they can receive targeted instruction and support that meets their individual needs.

Adaptive learning can also help to improve student outcomes. By providing students with targeted instruction and support, adaptive learning can help to close achievement gaps and improve student performance. Adaptive learning can also help to develop critical thinking skills, problem-solving skills, and the ability to apply knowledge to real-world situations.

However, there are also some challenges associated with adaptive learning. One of the biggest challenges is ensuring that the technology and algorithms are accurate and effective. This requires ongoing monitoring and evaluation to ensure that the technology is providing accurate assessments and recommendations.

Another challenge is ensuring that students are receiving adequate support and guidance. While adaptive learning can be highly engaging and motivating, students may need additional support from teachers and mentors to ensure that they are on track and receiving the support they need.

In conclusion, adaptive learning is an important component of the personalization approach to education that is a key feature of Education 5.0. Adaptive learning enables students to receive personalized instruction and support that meets their individual needs, and it can help to improve student outcomes and prepare students for success in this fast changing world. There are some challenges associated with adaptive learning, but when implemented effectively, it can be a powerful tool for improving student learning and engagement [101,102].

13.3.2.1.3 Flexible scheduling

Flexible scheduling is another important component of the personalization approach to education. Flexible scheduling recognizes that students have different needs and schedules and that traditional schedules may not be appropriate or effective for all students.

Flexible scheduling involves the use of a variety of scheduling options, including online learning, blended learning, and self-paced learning. These options allow students to work at their own pace and in their own way and they can help to accommodate different learning styles and preferences.

One of the key benefits of flexible scheduling is that it can help to increase student engagement and motivation. By providing students with more control over their learning environment and schedule, flexible scheduling can help to increase student ownership of their learning and their commitment to academic success.

Flexible scheduling can also help to improve student outcomes. By providing students with targeted instruction and support that meets their individual needs, flexible scheduling can help to close achievement gaps and improve student performance. Flexible scheduling can also help to develop critical thinking skills, problem-solving skills, and the ability to apply knowledge to real-world situations.

However, there are also some challenges associated with flexible scheduling. One of the biggest challenges is ensuring that students receive adequate support and guidance. While flexible scheduling can be highly engaging and motivating, students may need additional support from teachers and mentors to ensure that they are on track and receiving the support they need.

Another challenge is ensuring that assessments are aligned with learning objectives and standards, and that they are fair and consistent across different scheduling options. This requires ongoing monitoring and evaluation to ensure that assessments are accurate and effective, and that students are receiving appropriate feedback and support.

In conclusion, flexible scheduling is a critical component of the personalization approach to education that is a key feature of Education 5.0. Flexible scheduling enables students to receive instruction and support that meets their individual needs and schedules, and it can help to improve student outcomes and prepare students for success in fast changing world. While there are some challenges associated with flexible scheduling, when implemented effectively, it could be a powerful tool for improving student learning and engagement [103–105].

13.3.2.1.4 Personalized assessments

Personalized assessments are an important component of the personalization approach to education. Personalized assessments refer to the use of technology and data analytics to tailor assessments to the needs and preferences of individual students.

Personalized assessments involve the use of adaptive assessments, which use algorithms and analytics to adjust the difficulty of questions and tasks based on a student's performance. This ensures that students are being challenged appropriately and receiving assessments that are aligned with their level of understanding.

Personalized assessments can also involve the use of alternate assessments such as project-based assessments, portfolios, and performance tasks. These assessments can provide a more comprehensive view of a student's knowledge and skills and can help to better align assessments with learning objectives and standards.

One of the key benefits of personalized assessments is that they can provide students with more accurate and meaningful feedback on their progress. By tailoring assessments to the needs and preferences of individual students, personalized assessments can provide more targeted feedback that can help students identify areas where they need additional support and guidance.

Personalized assessments can also help to improve student outcomes. By providing students with more accurate and meaningful feedback on their progress, personalized assessments can help to close achievement gaps and improve student performance. Personalized assessments can also help to develop critical thinking skills, problem-solving skills, and the ability to apply knowledge to real-world situations.

However, there are also some challenges associated with personalized assessments. One of the biggest challenges is ensuring that the technology and algorithms are accurate and effective. This requires ongoing monitoring and evaluation to ensure that the technology is providing accurate assessments and recommendations.

Another challenge is ensuring that assessments are aligned with learning objectives and standards, and that they are fair and consistent across different students and learning paths. This requires ongoing monitoring and evaluation to ensure that assessments are accurate and effective, and that students are receiving appropriate feedback and support.

In conclusion, 'personalized assessments' is an important component of the personalization approach to education that is a key feature of Education 5.0. Personalized assessments can provide students with more accurate and meaningful feedback on their progress and they can help to improve student outcomes and prepare students for success in a rapidly changing world. While there are some challenges associated with personalized assessments, when implemented effectively, they can be a powerful tool for improving student learning and engagement [106,107].

Personalisation in Education 5.0 is designed to help learners achieve their full potential by providing them with the support, guidance, and resources they need to succeed in their learning journey.

13.3.2.2 Collaboration

Collaboration is another important feature of Education 5.0, which emphasizes the importance of working together to solve real-world problems and create innovative solutions and achieve learning goals. Collaboration can be fostered through project-based learning and other experiential learning approaches, which require students to work in teams to complete tasks and achieve learning objectives.

In Education 5.0, collaboration can take many different forms, such as peer-to-peer collaboration, group projects, and team-based learning activities.

Here are some ways in which collaboration is implemented in Education 5.0:

1. Peer-to-peer learning: Education 5.0 encourages learners to work together and learn from each other. This can include peer-to-peer mentoring, where more experienced learners provide guidance and support to their peers.

2. Group projects: Education 5.0 promotes collaborative learning through group projects and team-based assignments. This allows learners to develop their communication, leadership, and problem-solving skills, while also learning from the diverse perspectives of their peers.
3. Community engagement: Education 5.0 encourages learners to engage with their local and global communities, and to work together to address real-world problems. This can include service learning projects, community outreach programs, and global partnerships.
4. Virtual collaboration: Education 5.0 leverages technology to facilitate virtual collaboration and teamwork, allowing learners to work together regardless of their location or time zone. This can include online discussion forums, virtual group meetings, and collaborative document editing tools.

13.3.2.2.1 Peer-to-peer learning

Peer-to-peer learning is an important component of the collaboration approach to education that is another key feature of Education 5.0. Peer-to-peer learning refers to the use of collaborative learning strategies that involve students working together to solve problems, complete tasks and develop new knowledge and skills.

Peer-to-peer learning involves the use of small group activities, collaborative projects, and peer feedback and evaluation. These activities can help to improve communication skills, problem-solving skills, and teamwork skills and they can help create a more supportive and engaging learning environment.

One of the key benefits of peer-to-peer learning is that it can help to increase student engagement and motivation. By providing students with more opportunities to collaborate and work together, peer-to-peer learning can help to increase student ownership of their learning and their commitment to academic success.

Peer-to-peer learning can also help improve student outcomes. By providing students with more opportunities to work together and share their knowledge and skills, peer-to-peer learning can help close achievement gaps and improve student performance. Peer-to-peer learning can also help to develop critical thinking skills, problem-solving skills, and the ability to apply knowledge to real-world situations.

However, there are also some challenges associated with peer-to-peer learning. One of the biggest challenges is ensuring that students are receiving adequate support and guidance. While peer-to-peer learning can be highly engaging and motivating, students may need additional support from teachers and mentors to ensure that they are on track and receiving the support they need.

Another challenge is ensuring that assessments are fair and consistent across different students and learning paths. This requires ongoing monitoring and evaluation to ensure that assessments are accurate and effective, and that students are receiving appropriate feedback and support.

In conclusion, peer-to-peer learning is an important component of the collaboration approach to education that is a key feature of Education 5.0. Peer-to-peer learning can provide students with more opportunities to collaborate and work together and it can help improve student outcomes and prepare students for success in a fast changing world. While there are some challenges associated with peer-to-peer learning, when implemented effectively, it can be a powerful tool for improving student learning and engagement [108].

13.3.2.2.2 Group projects

'Group projects' is another important component of the collaboration approach to education. Group projects refer to the use of collaborative learning strategies that involve students working together in small groups to complete tasks, solve problems, and develop new knowledge and skills.

Group projects involve a range of activities including brainstorming, planning, research, and presentation of findings. These activities can help improve communication skills, problem-solving skills, and teamwork skills, and they can help create a more supportive and engaging learning environment.

One of the key benefits of group projects is that they can help increase student engagement and motivation. By providing students with more opportunities to work together and share their knowledge and skills, group projects can help to increase student ownership of their learning and their commitment to academic success.

Group projects can also help in improving student outcomes. By providing students with more opportunities to work together and share their knowledge and skills, group projects can help to close achievement gaps and improve student performance. Group projects can also help to develop critical thinking skills, problem-solving skills, and the ability to apply knowledge to real-world situations.

However, there are also some challenges associated with group projects. One of the biggest challenges is ensuring that students are receiving adequate support and guidance. While group projects can be highly engaging and motivating, students may need additional support from teachers and mentors to ensure that they are on track and receiving the support they need.

Another challenge is ensuring that group projects are fair and equitable. This requires careful planning and monitoring to ensure that each group is provided with equal resources and opportunities, and that each member of the group is given the opportunity to contribute and learn.

In conclusion, group projects are an important component of the collaboration approach to education. Group projects can provide students with more opportunities to collaborate and work together and they can help in improving student outcomes and prepare students for success in a rapidly changing world. While there are some challenges associated with group projects, when implemented effectively they can be a powerful tool for improving student learning and engagement [109].

13.3.2.2.3 Community engagement

Community engagement is an important component of the collaboration approach to education. Community engagement refers to the involvement of community members, organizations, and resources in the educational process to support student learning and development.

Community engagement involves a range of activities including service learning, internships, community-based research, and partnerships with local businesses and organizations. These activities can help to improve communication skills, problem-solving skills, and teamwork skills. They can also help in creating a more supportive and engaging learning environment.

One of the key benefits of community engagement is that it can help to increase student engagement and motivation. By providing students with more opportunities to work with community members and organizations, community engagement can help to increase student ownership of their learning and their commitment to academic success.

Community engagement can also help in improving student outcomes. By providing students with more opportunities to work with community members and organizations, community engagement can help to close achievement gaps and improve student performance. Community engagement can also help to develop critical thinking skills, problem-solving skills, and the ability to apply knowledge to real-world situations.

However, there are also some challenges associated with community engagement. One of the biggest challenges is ensuring that community engagement is integrated into the curriculum in a meaningful and effective way. This requires careful planning and monitoring to ensure that community engagement activities are aligned with learning objectives and standards and that they are providing students with the support and guidance they need.

Another challenge is ensuring that community engagement is equitable and inclusive. This requires ongoing monitoring and evaluation to ensure that community engagement activities are accessible to all students regardless of their background or circumstances and that they are providing students with the support and resources they need to succeed.

In conclusion, community engagement is an important component of the collaboration approach to education that is another key feature of

Education 5.0. Community engagement can provide students with more opportunities to work with community members and organizations and it can help to improve student outcomes and prepare students for success in a fast changing world. While there are some challenges associated with community engagement, when implemented effectively it can be a powerful tool for improving student learning and engagement [110,111].

13.3.2.2.4 Virtual collaboration

Virtual collaboration is another important component of the collaboration approach to education that is a key feature of Education 5.0. Virtual collaboration refers to the use of online tools and technologies to facilitate collaboration and communication between students and teachers in virtual environments.

Virtual collaboration involves a range of activities including online discussions, group projects, virtual classrooms, and online peer feedback and evaluation. These activities can help to improve communication skills, problem-solving skills, and teamwork skills, and they can help to create a more supportive and engaging learning environment.

One of the key benefits of virtual collaboration is that it can help to increase student engagement and motivation. By providing students with more opportunities to collaborate and communicate online, virtual collaboration can help to increase student ownership of their learning and their commitment to academic success.

Virtual collaboration can also help to improve student outcomes. By providing students with more opportunities to collaborate and communicate online, virtual collaboration can help to close achievement gaps and improve student performance. Virtual collaboration can also help to develop critical thinking skills, problem-solving skills, and the ability to apply knowledge to real-world situations.

However, there are also some challenges associated with virtual collaboration. One of the biggest challenges is ensuring that students are receiving adequate support and guidance. While virtual collaboration can be highly engaging and motivating, students may need additional support from teachers and mentors to ensure that they are on track and receiving the support they need.

Another challenge is ensuring that virtual collaboration is effective and meaningful. This requires careful planning and monitoring to ensure that virtual collaboration activities are aligned with learning objectives and standards, and that they are providing students with the support and guidance they need to succeed.

In conclusion, virtual collaboration is an important component of the collaboration approach to education that is a key feature of Education 5.0.

Virtual collaboration can provide students with more opportunities to collaborate and communicate online, and it can help improve student outcomes and prepare students for success in a rapidly changing world. While there are some challenges associated with virtual collaboration, when implemented effectively, it can be a powerful tool for improving student learning and engagement [112,113].

Hence, collaboration in Education 5.0 is designed to promote a culture of teamwork, communication, and mutual support, where learners can develop the skills and competencies they need to succeed in a complex and interconnected world.

13.3.2.3 Technology

Technology is a critical element of Education 5.0, which recognizes the transformative power of digital technologies in creating personalized, collaborative, and experiential learning experiences. Technology can be used to facilitate personalized learning, provide real-time feedback and assessment, and connect students to a global community of learners.

Technology is a key component of Education 5.0, which leverages the latest digital tools and platforms to support learning and enhance the educational experience. In Education 5.0, technology is used to facilitate personalized and adaptive learning, improve collaboration and communication, and provide learners with access to a wealth of learning resources.

Here are some ways in which technology is used in Education 5.0:

1. Online learning platforms: Education 5.0 uses online learning platforms to deliver courses and educational content to learners around the world. These platforms can include learning management systems (LMS), massive open online courses (MOOCs), and educational apps.
2. Virtual and augmented reality: Education 5.0 leverages virtual and augmented reality to create immersive and interactive learning experiences. This can include virtual field trips, simulations, and augmented reality games.
3. Artificial intelligence: Education 5.0 uses artificial intelligence (AI) to provide personalized and adaptive learning experiences. AI can analyze learner data and provide personalized recommendations for learning activities and resources.
4. Cloud computing: Education 5.0 uses cloud computing to provide learners with access to a vast array of learning resources, including digital textbooks, educational videos, and online databases.
5. Social media: Education 5.0 uses social media to facilitate collaboration and communication among learners and instructors. Social media platforms can be used for online discussions, collaborative projects, and peer-to-peer mentoring.

13.3.2.3.1 Online learning platforms

Online learning platforms are an important component of the technology approach to education that is a key feature of Education 5.0. Online learning platforms refer to the use of web-based tools and technologies to deliver educational content and resources to students and teachers in virtual environments.

Online learning platforms involve a range of activities, including online lectures, digital textbooks, multimedia presentations, interactive quizzes, and online discussion forums. These activities can help to improve access to educational resources, promote active learning and engagement, and support personalized learning paths for students.

One of the key benefits of online learning platforms is that they can help to increase access to educational opportunities. By providing students with access to educational resources online, online learning platforms can help to close the digital divide and provide equal access to educational opportunities for all students regardless of their location or background.

Online learning platforms can also help to improve student outcomes. By providing students with more opportunities for personalized learning, online learning platforms can help to close achievement gaps and improve student performance. Online learning platforms can also help to develop critical thinking skills, problem-solving skills, and the ability to apply knowledge to real-world situations.

However, there are also some challenges associated with online learning platforms. One of the biggest challenges is ensuring that students are receiving adequate support and guidance. While online learning platforms can be highly engaging and motivating, students may need additional support from teachers and mentors to ensure that they are on track and receiving the support they need.

Another challenge is ensuring that online learning platforms are effective and engaging. This requires careful planning and monitoring to ensure that online learning activities are aligned with learning objectives and standards and that they are providing students with the support and guidance they need to succeed.

In conclusion, online learning platforms are an important component of the technology approach to education. Online learning platforms can provide students with access to educational resources online, promote active learning and engagement, and support personalized learning paths for students. While there are some challenges associated with online learning platforms, when implemented effectively they can be a powerful tool for improving student learning and engagement in the 21st century [114,115].

13.3.2.3.2 Virtual and augmented reality

Virtual and augmented reality (VR and AR) are important components of the technology approach to education that is a key feature of Education 5.0. VR and AR technologies refer to the use of digital tools and technologies to create immersive learning environments that can help enhance student engagement and improve learning outcomes.

Virtual reality involves the use of computer-generated simulations to create a realistic and immersive learning environment. Virtual reality can help to create engaging and interactive learning experiences that can help to improve student engagement and learning outcomes. For example, virtual reality can be used to simulate complex scientific phenomena or historical events, allowing students to explore and interact with these phenomena in a more engaging and meaningful way.

Augmented reality involves the use of digital tools and technologies to overlay digital information onto the real world. Augmented reality can help in enhancing the learning experience by providing students with additional information and context that can help to support learning outcomes. For example, augmented reality can be used to overlay information about a historical site or a scientific experiment onto the real world, allowing students to explore and interact with these phenomena in a more engaging and meaningful way.

One of the key benefits of VR and AR technologies is that they can help to increase student engagement and motivation. By providing students with more engaging and interactive learning experiences, VR and AR technologies can help to increase student ownership of their learning and their commitment to academic success.

VR and AR technologies can also help in improving student outcomes. By providing students with more opportunities to explore and interact with complex phenomena in a more engaging and meaningful way, VR and AR technologies can help to improve student performance and prepare students for success in a rapidly changing world.

However, there are also some challenges associated with VR and AR technologies. One of the biggest challenges is ensuring that these technologies are integrated into the curriculum in a meaningful and effective way. This requires careful planning and monitoring to ensure that VR and AR activities are aligned with learning objectives and standards, and that they are providing students with the support and guidance they need.

Another challenge is ensuring that VR and AR technologies are accessible to all students. This requires ongoing monitoring and evaluation to ensure that VR and AR activities are accessible to all students, regardless of their background or circumstances and that they are providing students with the support and resources they need to succeed.

In conclusion, VR and AR technologies are important components of the technology approach to education that is a key feature of Education 5.0. VR and AR technologies can provide students with engaging and interactive learning experiences, enhance learning outcomes, and prepare students for success in a rapidly changing world. While there are some challenges associated with VR and AR technologies, when implemented effectively, they can be a powerful tool for improving student learning and engagement [116,117].

13.3.2.3.3 Artificial intelligence

Artificial intelligence (AI) is an important component of the technology approach to education that is a key feature of Education 5.0. AI technologies refer to the use of digital tools and technologies to simulate human intelligence and behavior, allowing machines to learn, reason and make decisions.

AI technologies can be used in a range of educational contexts including personalized learning, data analytics, and automated assessment. AI technologies can help to improve student outcomes by providing students with more personalized and adaptive learning experiences and by providing teachers with more accurate and timely feedback on student performance.

One of the key benefits of AI technologies is that they can help to increase student engagement and motivation. By providing students with more personalized and adaptive learning experiences, AI technologies can help to increase student ownership of their learning and their commitment to academic success.

AI technologies can also help to improve student outcomes. By providing students with more opportunities for personalized and adaptive learning, AI technologies can help to close achievement gaps and improve student performance. AI technologies can also help to develop critical thinking skills, problem-solving skills, and the ability to apply knowledge to real-world situations.

However, there are also some challenges associated with AI technologies. One of the biggest challenges is ensuring that these technologies are integrated into the curriculum in a meaningful and effective way. This requires careful planning and monitoring to ensure that AI activities are aligned with learning objectives and standards, and that they are providing students with the support and guidance they need.

Another challenge is ensuring that AI technologies are accessible to all students. This requires ongoing monitoring and evaluation to ensure that AI activities are accessible to all students, regardless of their background or circumstances, and that they are providing students with the support and resources they need to succeed.

In conclusion, AI technologies are an important component of the technology approach to education that is a key feature of Education 5.0. AI technologies can provide students with personalized and adaptive learning experiences, enhance learning outcomes, and prepare students for success in a rapidly changing world. While there are some challenges associated with AI technologies, when implemented effectively, they can be a powerful tool for improving student learning and engagement [118,119].

13.3.2.3.4 Cloud computing

Cloud computing is an important component of the technology approach to education that is a key feature of Education 5.0. Cloud computing refers to the use of remote servers, networks, and storage devices to store, manage, and process data and applications.

Cloud computing can be used in a range of educational contexts, including personalized learning, data analytics, and collaborative learning. Cloud computing can help to improve student outcomes by providing students with access to educational resources and tools from anywhere, at any time.

One of the key benefits of cloud computing is that it can help to increase access to educational resources. By providing students with access to educational resources and tools online, cloud computing can help to close the digital divide and provide equal access to educational opportunities for all students, regardless of their location or background.

Cloud computing can also help to improve student outcomes. By providing students with more opportunities for personalized and collaborative learning, cloud computing can help to improve student performance and prepare students for success in a rapidly changing world.

However, there are also some challenges associated with cloud computing. One of the biggest challenges is ensuring that cloud-based resources and tools are secure and protected against data breaches and cyber-attacks. This requires ongoing monitoring and evaluation to ensure that cloud-based resources and tools are secure and protected against unauthorized access or use.

Another challenge is ensuring that cloud-based resources and tools are accessible to all students. This requires ongoing monitoring and evaluation to ensure that cloud-based resources and tools are accessible to all students, regardless of their background or circumstances, and that they are providing students with the support and resources they need to succeed.

In conclusion, cloud computing is an important component of the technology approach to education that is a key feature of Education 5.0. Cloud computing can provide students with access to educational resources and tools from anywhere, at any time, enhance learning outcomes, and prepare students for success in a rapidly changing world. While there are some challenges associated with cloud computing, when implemented effectively, it can be a powerful tool for improving student learning and engagement [120,121].

13.3.2.3.5 Social media

Social media is an important component of the technology approach to education that is a key feature of Education 5.0. Social media refers to the use of online platforms and tools that allow individuals to create, share, and exchange information and ideas with others.

Social media can be used in a range of educational contexts including personalized learning, collaborative learning, and data analytics. Social media can help to improve student outcomes by providing students with access to educational resources, facilitating collaboration and communication, and promoting engagement and participation.

One of the key benefits of social media is that it can help to increase student engagement and participation. By providing students with opportunities to interact and engage with peers and teachers, social media can help to create a more collaborative and interactive learning environment.

Social media can also help to improve student outcomes. By providing students with more opportunities for collaborative learning, social media can help to improve student performance and prepare students for success in a rapidly changing world. Social media can also help to develop critical thinking skills, problem-solving skills, and the ability to apply knowledge to real-world situations.

However, there are also some challenges associated with social media. One of the biggest challenges is ensuring that social media is used in a safe and responsible way. This requires ongoing monitoring and evaluation to ensure that social media is used in a way that is respectful of others and that protects student privacy and safety.

Another challenge is ensuring that social media is accessible to all students. This requires ongoing monitoring and evaluation to ensure that social media is accessible to all students, regardless of their background or circumstances, and that they are providing students with the support and resources they need to succeed.

In conclusion, social media is an important component of the technology approach to education. Social media can provide students with access to educational resources, facilitate collaboration and communication, enhance learning outcomes, and prepare students for success in a rapidly changing world. While there are some challenges associated with social media, when implemented effectively, it can be a powerful tool for improving student learning and engagement [122,123].

Technology in Education 5.0 is designed to enhance the learning experience, making it more engaging, interactive, and personalized. It provides learners with access to a wealth of educational resources and opportunities regardless of their location or background.

13.3.2.4 Experiential learning

Experiential learning is a core component of Education 5.0, which recognizes that students learn best through hands-on experiences and real-world applications of knowledge and skills. Experiential learning can be facilitated through activities such as internships, service learning, and simulations, which allow students to apply their learning in real-world contexts and develop practical skills.

Experiential learning is an important feature of Education 5.0, which emphasizes the importance of hands-on real-world learning experiences. In Education 5.0, experiential learning is designed to help learners apply their knowledge and skills to practical situations and to develop a deeper understanding of the world around them.

Here are some ways in which experiential learning is implemented in Education 5.0 [14,124]:

1. Internships and apprenticeships: Education 5.0 provides learners with opportunities to gain practical experience through internships, apprenticeships, and other work-based learning programs. This allows learners to apply their classroom learning to real-world situations and to develop professional skills and competencies.
2. Service learning: Education 5.0 encourages learners to engage in service learning, where they apply their knowledge and skills to address real-world problems and make a positive impact in their communities.
3. Project-based learning: Education 5.0 promotes project-based learning, where learners work on real-world projects that are relevant to their interests and career aspirations. This allows learners to develop critical thinking, problem-solving, and collaboration skills, while also applying their knowledge to practical situations.
4. Study abroad programs: Education 5.0 provides learners with opportunities to study abroad, where they can gain new perspectives, learn about different cultures, and develop global competencies.
5. Experiential simulations: Education 5.0 leverages technology to create experiential simulations such as virtual reality and augmented reality that allow learners to practice real-world skills and scenarios in a safe and controlled environment.

13.3.2.4.1 Internships and apprenticeships

Internships and apprenticeships are important components of the experiential learning approach to education that is a key feature of Education 5.0. Experiential learning refers to the use of real-world experiences and practical applications to enhance student learning and understanding.

Internships and apprenticeships can be used in a range of educational contexts including career and technical education, workforce development, and higher education. Internships and apprenticeships can help to improve student outcomes by providing students with opportunities to apply knowledge and skills in real-world settings, develop new skills, and also build professional networks.

One of the key benefits of internships and apprenticeships is that they can help to increase student engagement and motivation. By providing students with opportunities to apply knowledge and skills in real-world settings, internships and apprenticeships can help to increase student ownership of their learning and their commitment to academic success.

Internships and apprenticeships can also help to improve student outcomes. By providing students with more opportunities for hands-on learning and practical application of knowledge and skills, internships and apprenticeships can help to close achievement gaps and improve student performance. Internships and apprenticeships can also help to develop critical thinking skills, problem-solving skills, and the ability to apply knowledge to real-world situations.

However, there are also some challenges associated with internships and apprenticeships. One of the biggest challenges is ensuring that these experiences are designed in a meaningful and effective way. This requires careful planning and monitoring to ensure that internships and apprenticeships are aligned with learning objectives and standards and that they are providing students with the support and guidance they need.

Another challenge is ensuring that internships and apprenticeships are accessible to all students. This requires ongoing monitoring and evaluation to ensure that internships and apprenticeships are accessible to all students, regardless of their background or circumstances, and that they are providing students with the support and resources they need to succeed.

In conclusion, internships and apprenticeships are important components of the experiential learning approach to education that is a key feature of Education 5.0. Internships and apprenticeships can provide students with opportunities to apply knowledge and skills in real-world settings, develop new skills, and build professional networks. While there are some challenges associated with internships and apprenticeships, when implemented effectively, they can be a powerful tool for improving student learning and engagement.

13.3.2.4.2 Service learning

Service learning is an important component of the experiential learning approach to education that is a key feature of Education 5.0. Service learning refers to the use of community service and volunteer work as a way to enhance student learning and understanding.

Service learning can be used in a range of educational contexts, including K-12 education, higher education, and workforce development. Service learning can help to improve student outcomes by providing students with opportunities to apply knowledge and skills in real-world settings, develop empathy and social awareness, and build relationships with community members.

One of the key benefits of service learning is that it can help to increase student engagement and motivation. By providing students with opportunities to make a positive impact in their communities, service learning can help to increase student ownership of their learning and their commitment to academic success.

Service learning can also help to improve student outcomes. By providing students with more opportunities for hands-on learning and practical application of knowledge and skills, service learning can help to close achievement gaps and improve student performance. Service learning can also help to develop critical thinking skills, problem-solving skills, and the ability to apply knowledge to real-world situations.

However, there are also some challenges associated with service learning. One of the biggest challenges is ensuring that service learning experiences are designed in a meaningful and effective way. This requires careful planning and monitoring to ensure that service learning experiences are aligned with learning objectives and standards, and that they are providing students with the support and guidance they need.

Another challenge is ensuring that service learning experiences are accessible to all students. This requires ongoing monitoring and evaluation to ensure that service learning experiences are accessible to all students, regardless of their background or circumstances, and that they are providing students with the support and resources they need to succeed.

In conclusion, service learning is an important component of the experiential learning approach to education that is a key feature of Education 5.0. Service learning can provide students with opportunities to apply knowledge and skills in real-world settings, develop empathy and social awareness, and build relationships with community members. While there are some challenges associated with service learning, when implemented effectively, it can be a powerful tool for improving student learning and engagement.

13.3.2.4.3 Project-based learning

Project-based learning is an important component of the experiential learning approach to education that is a key feature of Education 5.0. Project-based learning refers to the use of real-world projects and problem-solving as a way to enhance student learning and understanding.

Project-based learning can be used in a range of educational contexts, including K-12 education, higher education, and workforce development. Project-based learning can help to improve student outcomes by providing students with opportunities to apply knowledge and skills in real-world settings, develop new skills, and build teamwork and collaboration.

One of the key benefits of project-based learning is that it can help to increase student engagement and motivation. By providing students with opportunities to work on meaningful and relevant projects, project-based learning can help to increase student ownership of their learning and their commitment to academic success.

Project-based learning can also help to improve student outcomes. By providing students with more opportunities for hands-on learning and practical application of knowledge and skills, project-based learning can help to close achievement gaps and improve student performance. Project-based learning can also help to develop critical thinking skills, problem-solving skills, and the ability to apply knowledge to real-world situations.

However, there are also some challenges associated with project-based learning. One of the biggest challenges is ensuring that project-based learning experiences are designed in a meaningful and effective way. This requires careful planning and monitoring to ensure that projects are aligned with learning objectives and standards, and that they are providing students with the support and guidance they need.

Another challenge is ensuring that project-based learning experiences are accessible to all students. This requires ongoing monitoring and evaluation to ensure that project-based learning experiences are accessible to all students, regardless of their background or circumstances, and that they are providing students with the support and resources they need to succeed.

In conclusion, project-based learning is an important component of the experiential learning approach to education that is a key feature of Education 5.0. Project-based learning can provide students with opportunities to apply knowledge and skills in real-world settings, develop new skills, and build teamwork and collaboration. While there are some challenges associated with project-based learning, when implemented effectively, it can be a powerful tool for improving student learning and engagement.

13.3.2.4.4 Study abroad programs

Study abroad programs are an important component of the experiential learning approach to education that is a key feature of Education 5.0. Study abroad programs refer to the use of travel and study in foreign countries as a way to enhance student learning and understanding.

Study abroad programs can be used in a range of educational contexts, including higher education and workforce development. Study abroad programs can help to improve student outcomes by providing students with opportunities to immerse themselves in new cultures, learn new languages, and gain a global perspective on their field of study.

One of the key benefits of study abroad programs is that they can help to increase student engagement and motivation. By providing students with opportunities to explore new places and cultures, study abroad programs can help to increase student ownership of their learning and their commitment to academic success.

Study abroad programs can also help to improve student outcomes. By providing students with more opportunities for hands-on learning and practical application of knowledge and skills, study abroad programs can help to close achievement gaps and improve student performance. Study abroad programs can also help to develop critical thinking skills, problem-solving skills, and the ability to apply knowledge to real-world situations.

However, there are also some challenges associated with study abroad programs. One of the biggest challenges is ensuring that study abroad programs are designed in a meaningful and effective way. This requires careful planning and monitoring to ensure that study abroad experiences are aligned with learning objectives and standards and that they are providing students with the support and guidance they need.

Another challenge is ensuring that study abroad experiences are accessible to all students. This requires ongoing monitoring and evaluation to ensure that study abroad experiences are accessible to all students regardless of their background or circumstances and that they are providing students with the support and resources they need to succeed.

In conclusion, study abroad programs are an important component of the experiential learning approach to education that is a key feature of Education 5.0. Study abroad programs can provide students with opportunities to immerse themselves in new cultures, learn new languages, and gain a global perspective on their field of study. While there are some challenges associated with study abroad programs, when implemented effectively, they can be a powerful tool for improving student learning and engagement.

13.3.2.4.5 Experiential simulations

Experiential simulations are an important component of the experiential learning approach to education that is a key feature of Education 5.0. Experiential simulations refer to the use of simulated environments and scenarios as a way to enhance student learning and understanding.

Experiential simulations can be used in a range of educational contexts, including K-12 education, higher education, and workforce development. Experiential simulations can help to improve student outcomes by providing students with opportunities to apply knowledge and skills in simulated real-world settings, develop new skills, and build teamwork and collaboration.

One of the key benefits of experiential simulations is that they can help to increase student engagement and motivation. By providing students with opportunities to work in realistic simulated environments and scenarios, experiential simulations can help to increase student ownership of their learning and their commitment to academic success.

Experiential simulations can also help to improve student outcomes. By providing students with more opportunities for hands-on learning and practical application of knowledge and skills, experiential simulations can help to close achievement gaps and improve student performance. Experiential simulations can also help to develop critical thinking skills, problem-solving skills, and the ability to apply knowledge to real-world situations.

However, there are also some challenges associated with experiential simulations. One of the biggest challenges is ensuring that experiential simulations are designed in a meaningful and effective way. This requires careful planning and monitoring to ensure that simulations are aligned with learning objectives and standards and that they are providing students with the support and guidance they need.

Another challenge is ensuring that experiential simulations are accessible to all students. This requires ongoing monitoring and evaluation to ensure that experiential simulations are accessible to all students, regardless of their background or circumstances, and that they are providing students with the support and resources they need to succeed.

In conclusion, experiential simulations are an important component of the experiential learning approach to education that is a key feature of Education 5.0. Experiential simulations can provide students with opportunities to apply knowledge and skills in simulated real-world settings, develop new skills, and build teamwork and collaboration. While there are some challenges associated with experiential simulations, when implemented effectively, they can be a powerful tool for improving student learning and engagement.

And so, experiential learning in Education 5.0 is designed to provide learners with opportunities to apply their knowledge and skills to practical situations and to develop the professional skills and competencies they need to succeed in their careers and personal lives.

13.3.2.5 Globalisation

Globalization is another key feature of Education 5.0, which recognizes the interconnectedness of the world and the need for students to develop global competencies and perspectives. Globalization can be facilitated through international study programs, language learning and virtual exchange programs, which allow students to connect with peers from around the world and develop a global mindset.

Globalisation is a key feature of Education 5.0, which emphasizes the importance of preparing learners to succeed in an increasingly interconnected and globalized world. In Education 5.0, globalisation means that learners gain a deep understanding of different cultures, perspectives, and languages and are equipped with the skills and competencies they need to work and communicate effectively in a global context.

Here are some ways in which globalisation is implemented in Education 5.0:

1. Multicultural education: Education 5.0 promotes multicultural education, where learners gain an understanding of different cultures, histories, and perspectives. This can include courses and programs that focus on global issues, such as human rights, climate change, and sustainable development.
2. Language learning: Education 5.0 encourages learners to learn multiple languages, and to develop the language skills they need to communicate effectively in a global context.

3. International study programs: Education 5.0 provides learners with opportunities to study abroad, where they can gain new perspectives, learn about different cultures, and develop global competencies.
4. Global partnerships: Education 5.0 encourages educational institutions and organizations to form global partnerships, where they can collaborate on research, curriculum development, and student exchange programs.
5. Online global learning communities: Education 5.0 leverages technology to create online global learning communities, where learners can connect with peers and instructors from around the world and collaborate on projects and learning activities.

13.3.2.5.1 Multicultural education

Multicultural education is an important component of the globalization aspect of Education 5.0. It refers to the integration of diverse cultural perspectives, experiences, and values into the learning process, with the goal of fostering respect, empathy, and understanding for people from different backgrounds.

Multicultural education is critical in today's globalized world, where people from different cultural backgrounds are increasingly interacting with each other in both personal and professional contexts. Multicultural education can help to promote intercultural competence and communication skills, which are essential for success in a globalized workforce.

There are several strategies that can be used to incorporate multicultural education into the learning process. One strategy is to use culturally responsive teaching, which involves using teaching methods and materials that are culturally relevant and responsive to the diverse backgrounds and experiences of students.

Another strategy is to use service learning, which involves engaging students in community-based projects that address social and cultural issues. This can help to foster a sense of social responsibility and empathy for people from different cultural backgrounds.

Finally, using technology to connect with people from different cultural backgrounds can be another effective strategy. For example, video conferencing and social media can be used to facilitate intercultural communication and collaboration among students from different parts of the world.

However, implementing multicultural education can also present challenges. One challenge is the need for teachers to have the necessary training and expertise to effectively integrate multicultural education into the learning process. This requires ongoing professional development and training to ensure that teachers have the necessary skills and knowledge to effectively incorporate multicultural education into their teaching practices.

Another challenge is the need to address issues of equity and access in multicultural education. This requires ensuring that all students have equal access to the resources and opportunities necessary to succeed in a multicultural learning environment.

In conclusion, multicultural education is an essential component of the globalization aspect of Education 5.0. Multicultural education can help to promote intercultural competence, communication skills, and social responsibility, which are essential for success in a globalized workforce. While implementing multicultural education can present challenges, when done effectively, it can be a powerful tool for promoting respect, empathy, and understanding for people from different cultural backgrounds [125–128].

13.3.2.5.2 Language learning

Language learning is an essential component of the globalization aspect of Education 5.0. With the world becoming more interconnected, the ability to communicate in multiple languages has become increasingly important for both personal and professional success.

Research has shown that learning a second language has numerous benefits, including improved cognitive function, increased cultural awareness, and better job prospects. Language learning can also help to facilitate intercultural communication and understanding, which is essential for success in a globalized world.

There are several strategies that can be used to promote language learning in the classroom. One effective strategy is to use immersion techniques, which involve creating an environment where the language being learned is the primary means of communication. This can help to accelerate language acquisition and improve language proficiency.

Another effective strategy is to incorporate technology into language learning. Technology can be used to provide students with access to authentic language resources, such as videos, podcasts, and news articles. Technology can also be used to facilitate language exchanges with students from other countries, which can help to promote intercultural communication and understanding.

However, implementing language learning can present challenges. One challenge is the need for teachers to have the necessary training and expertise to effectively teach a second language. This requires ongoing professional development and training to ensure that teachers have the necessary skills and knowledge to effectively teach a second language.

Another challenge is the need to address issues of equity and access in language learning. This requires ensuring that all students have equal access to the resources and opportunities necessary to succeed in language learning.

In conclusion, language learning is an essential component of the globalization aspect of Education 5.0. Language learning can help to promote cognitive function, cultural awareness, and intercultural communication and understanding, which are essential for success in a globalized world. While implementing language learning can present challenges, when done effectively, it can be a powerful tool for promoting language proficiency and intercultural communication and understanding [129–134].

13.3.2.5.3 International study programs

International study programs are an important component of the globalization aspect of Education 5.0. These programs provide students with the opportunity to study in a different country and immerse themselves in a new culture, while also gaining valuable academic and professional experience.

International study programs can take many forms, including study abroad programs, international internships, and international research programs. These programs can be a powerful tool for promoting intercultural communication and understanding, as well as for developing the skills and knowledge necessary for success in a globalized workforce.

Research has shown that international study programs can have a positive impact on student learning outcomes. For example, studies have found that students who participate in study abroad programs have higher GPAs are more likely to graduate on time and are more likely to be employed after graduation.

International study programs can also provide students with valuable cross-cultural skills and experiences, such as intercultural communication, adaptability, and problem-solving. These skills are highly valued by employers in a globalized workforce and can give students a competitive edge in their careers.

However, implementing international study programs can present challenges. One challenge is the need for adequate funding and resources to support international study programs. This requires ensuring that students have access to the financial and logistical resources necessary to participate in international study programs.

Another challenge is the need to ensure that international study programs are culturally sensitive and respectful. This requires ensuring that students are aware of cultural differences and are able to navigate cultural differences in a respectful and appropriate manner.

In conclusion, international study programs are an important component of the globalization aspect of Education 5.0. These programs provide students with valuable academic and professional experience, as well as valuable cross-cultural skills and experiences. While implementing international study programs can present challenges, when done effectively, they can be a powerful tool for promoting intercultural communication and understanding, as well as for developing the skills and knowledge necessary for success in a globalized workforce [135–138].

13.3.2.5.4 Global partnerships

Global partnerships are a critical aspect of the globalization aspect of Education 5.0. These partnerships involve collaborations between educational institutions, organizations, and individuals across international boundaries, with the aim of promoting intercultural communication and understanding, as well as advancing education and research in a global context.

Global partnerships can take many forms, including partnerships between universities, partnerships between educational institutions and non-governmental organizations, and partnerships between educational institutions and industry. These partnerships can be a powerful tool for promoting intercultural communication and understanding, as well as for advancing education and research in a globalized context.

Research has shown that global partnerships can have a positive impact on student learning outcomes. For example, studies have found that students who participate in international collaborations have higher levels of intercultural competence and are more likely to have positive attitudes toward other cultures.

Global partnerships can also provide students and educators with valuable opportunities for professional development, networking, and collaboration. These opportunities can help to facilitate the exchange of ideas and best practices across international boundaries and can lead to new insights and innovations in education and research.

However, implementing global partnerships can present challenges. One challenge is the need to navigate cultural differences and ensure that all partners are aware of cultural differences and are able to navigate them in a respectful and appropriate manner.

Another challenge is the need for adequate funding and resources to support global partnerships. This requires ensuring that all partners have access to the financial and logistical resources necessary to participate in global partnerships.

In conclusion, global partnerships are a critical aspect of the globalization aspect of Education 5.0. These partnerships can be a powerful tool for promoting intercultural communication and understanding, as well as for advancing education and research in a global context. While implementing global partnerships can present challenges, when done effectively, they can provide students and educators with valuable opportunities for professional development, networking, and collaboration, and can lead to new insights and innovations in education and research [139–142].

13.3.2.5.5 Online global learning communities

Online global learning communities are a crucial aspect of globalization in Education 5.0. These communities bring together students, educators, and professionals from around the world to collaborate and learn from each other. These communities are made possible by advancements in technology, which have made it easier than ever to connect with people from different parts of the world.

Online global learning communities offer several benefits. First, they provide students with the opportunity to learn from experts and peers from different cultural backgrounds. This exposure to different perspectives can help students develop a more nuanced understanding of the world and can prepare them for global citizenship.

Second, online global learning communities provide a platform for collaboration and networking. Students and professionals from different parts of the world can work together on projects, share ideas, and build relationships that can be valuable throughout their careers.

Finally, online global learning communities provide access to resources and expertise that might not otherwise be available. Students and educators can access lectures, research papers, and other materials from experts around the world, and can participate in discussions and forums where they can share ideas and ask questions.

However, there are challenges associated with online global learning communities. One challenge is the need to navigate cultural differences and ensure that all participants are aware of cultural differences and are able to navigate them in a respectful and appropriate manner. Another challenge is the need to ensure that all participants have access to the necessary technology and resources, which can be a barrier for students and educators in some parts of the world.

Despite these challenges, online global learning communities are an important tool for promoting globalization in Education 5.0. By providing students and educators with access to diverse perspectives, collaboration opportunities, and valuable resources and expertise, online global learning communities can help prepare students for success in a globalized world [143–146].

Globalization in Education 5.0 is designed to prepare learners to work and communicate effectively in a global context and to develop the skills and competencies they need to succeed in a complex and interconnected world.

Education 5.0 is a model of education that is designed to prepare learners for the challenges and opportunities of the 21st century. It emphasizes the importance of personalized and adaptive learning, collaboration and teamwork, experiential learning, globalization and multiculturalism, and technology-enabled learning, and encourages learners to be lifelong learners who are prepared to adapt to a rapidly changing world.

13.3.3 Life Long learning

Lifelong learning is the process of gaining knowledge and skills throughout one's life, from childhood to old age. It is based on the belief that learning is a continuous process that should never stop, and that everyone has the capacity to learn and develop throughout their life.

Lifelong learning is not limited to formal education, but can happen in a variety of settings, such as through work, travel, hobbies, and personal interests. It can also take many different forms, such as self-directed learning, online courses, workshops, apprenticeships, and mentoring.

The benefits of lifelong learning are numerous. It can help individuals stay competitive in the job market, improve their job prospects, and increase their earning potential. It can also lead to personal growth and fulfilment, as individuals pursue their passions and interests and gain a deeper understanding of themselves and the world around them.

Moreover, lifelong learning is increasingly important in today's rapidly changing world, where new technologies, industries, and job roles are constantly emerging. By embracing lifelong learning, individuals can stay up-to-date with the latest developments in their field, adapt to new challenges, and remain relevant in a constantly evolving job market.

Lifelong learning is the ongoing process of acquiring new knowledge, skills, and competencies throughout one's life. It is becoming increasingly important in today's rapidly changing world, where new technologies and globalization are constantly transforming the workplace and the economy. Education 5.0, with its focus on personalization, collaboration, and globalization, is well suited to support lifelong learning.

One of the key features of Education 5.0 is its emphasis on personalization. This includes customized learning paths, adaptive learning, and flexible scheduling, all of which can support lifelong learning. By allowing individuals to tailor their learning experiences to their specific needs and interests, Education 5.0 can help to ensure that individuals stay engaged and motivated to learn throughout their lives.

Another important feature of Education 5.0 is its emphasis on collaboration. Peer-to-peer learning, group projects, and community engagement can all provide opportunities for individuals to learn from and with others, and to build networks and relationships that can support lifelong learning. By fostering a sense of community and connection, Education 5.0 can help to create an environment in which lifelong learning is not only possible, but enjoyable and rewarding.

Finally, Education 5.0's focus on globalization can help to support lifelong learning by providing access to a diverse range of perspectives and expertise from around the world. Online global learning communities, international study programs, and global partnerships can all provide opportunities for individuals to expand their horizons and learn about different cultures, economies, and ways of life. By promoting a global perspective and encouraging individuals to be lifelong learners, Education 5.0 can help to ensure that individuals are equipped to succeed in a rapidly changing world.

However, there are also challenges associated with lifelong learning in Education 5.0. One challenge is the need for individuals to take responsibility for their own learning and development. In Education 5.0, learners are expected to be self-directed and self-motivated, which can be a challenge for some individuals. Another challenge is the need for individuals to be adaptable and flexible in their approach to learning, as new technologies and ways of working continue to emerge.

Despite these challenges, lifelong learning is becoming increasingly important in today's economy, and Education 5.0 is well suited to support it. By providing individuals with the tools and resources they need to continue learning and growing throughout their lives, Education 5.0 can help to ensure that individuals are equipped to succeed in a rapidly changing world [147–153].

Overall, lifelong learning is a valuable and essential practice for individuals who want to achieve personal and professional success, and thrive in an ever-changing world.

13.4 CHALLENGES AND SOLUTIONS

Despite the many benefits of Education 5.0, there are also several challenges that must be addressed in order to fully realize its potential. One major challenge is the digital divide, which limits access to technology and online learning resources for disadvantaged students. To address this challenge, schools, and governments must invest in digital infrastructure and provide equitable access to technology and internet connectivity for all students.

Another challenge is the need to train educators to effectively implement Education 5.0 approaches and technologies. Professional development opportunities need to be provided to help educators develop the skills and knowledge required to create personalized, collaborative and experiential learning experiences.

Finally, there is a need to develop new assessment models that can accurately measure the competencies and skills that are emphasized in Education 5.0. Traditional assessments such as standardized tests may not be adequate to assess the complex and diverse range of competencies that students are expected to develop in Education 5.0. New assessment models such as performance-based assessments and portfolios should be developed and implemented to accurately measure student learning and progress.

Let us discuss the above three challenges in detail in the following section.

13.4.1 Digital divide

The digital divide refers to the gap between those who have access to digital technologies and those who do not have that access. It is a significant challenge in the context of Education 5.0, as digital technologies are becoming increasingly important for learning and development.

One of the key challenges associated with the digital divide is unequal access to digital technologies. This can include access to high-speed internet, computers, and mobile devices. Individuals who do not have access to these technologies may be at a disadvantage when it comes to learning and development.

Another challenge is unequal digital literacy. Even if individuals have access to digital technologies, they may not have the skills and knowledge necessary to use them effectively for learning and development. This can include skills such as online research, digital collaboration, and data analysis.

To address these challenges, there are a number of potential solutions that can be implemented in the context of Education 5.0. One solution is to provide greater access to digital technologies, such as through initiatives to expand broadband internet access in underserved areas. Another solution is to provide digital skills training and support, such as through online tutorials, workshops, and mentoring programs.

In addition, there are a number of strategies that can be used to promote digital inclusion and reduce the digital divide. For example, educators can design learning materials and assessments that take into account different levels of digital literacy, and can provide feedback and support in helping learners develop their digital skills. Similarly, digital learning platforms can be designed to be accessible and inclusive, with features such as closed captioning, audio descriptions, and alternative text for images.

Ultimately, addressing the digital divide is essential for ensuring that Education 5.0 is accessible and equitable for all learners. By promoting greater access to digital technologies and digital literacy, and by designing learning experiences that are inclusive and supportive, we can help in ensuring that all individuals have the opportunity to learn and grow in the 21st century [154–159].

13.4.2 Training educators

As Education 5.0 continues to evolve, it is essential that educators are equipped with the skills and knowledge necessary to effectively integrate new technologies and teaching methodologies into their practice. However, training educators in the context of Education 5.0 is not without its challenges.

One challenge is the need for educators to develop new skills and competencies such as digital literacy, data analysis, and collaborative problem-solving. This can require significant time and resources, and may require educators to step outside of their comfort zones and embrace new/novel approaches to teaching and learning.

Another challenge is the need for ongoing professional development and support. As new technologies and methodologies continue to emerge, educators must be able to keep themselves up-to-date with the latest trends and best practices. This can require ongoing training, coaching, and mentoring, which can be difficult to provide in the context of busy classrooms and limited resources.

To address these challenges, there are a number of potential solutions that can be implemented in the context of Education 5.0. One solution is to provide targeted training and professional development opportunities for educators, such as workshops, webinars, and online courses. These opportunities can help educators in developing the skills and knowledge necessary to effectively integrate new technologies and methodologies into their practice.

Another solution is to provide ongoing support and coaching to educators through mentoring programs or peer-to-peer networks. This will ensure that educators have access to the resources and support they need to continue developing their skills and knowledge over time.

In addition, there are many other strategies that can be used to promote a culture of continuous learning and improvement among educators. For example, schools and educational institutions can create opportunities for educators to collaborate and share best practices, such as through online forums or regular meetings. Similarly, schools can encourage educators to engage in ongoing self-reflection and assessment, and provide support and feedback to help them identify areas for improvement and growth.

Ultimately, training educators is essential for ensuring that Education 5.0 is successful in delivering high-quality, student-centered learning experiences. By providing targeted training and support, and promoting a culture of continuous learning and improvement, we can help educators in developing the skills and competencies necessary to succeed in the 21st century classroom [160–165].

13.4.3 Developing new assessment models

As Education 5.0 continues to evolve, traditional assessment models may no longer be sufficient to measure the complex skill set and competencies that the students need to succeed in the 21st century. Developing new assessment models that are better suited to the needs of Education 5.0 presents a number of challenges, as well as potential solutions.

One challenge is the need to measure a broader range of skill set and competencies beyond traditional academic subjects. This may include skills such as creativity, critical thinking, collaboration, and problem-solving. Developing effective assessments for these skills can be difficult, and may require the use of alternative assessment methods such as performance tasks or portfolios.

Another challenge is the need to measure student progress over time, rather than simply at the end of a course in a semester/trimester or school year. This requires the development of ongoing assessment tools that can provide educators with real-time data on student learning, allowing them to adjust their teaching strategies as needed.

To address these challenges, there are a number of potential solutions that can be implemented in the context of Education 5.0. One solution is to develop new assessment models that are specifically designed to measure the skills and competencies that are most relevant to Education 5.0. This may involve the use of innovative assessment methods such as game-based assessments, simulations, or virtual reality environments.

Another solution is to incorporate ongoing assessment into the learning process, rather than treating it as a separate event. This can involve the use of formative assessment tools, such as quizzes, surveys, or feedback mechanisms, that allow educators to monitor student progress and adjust their teaching strategies in real time.

In addition, there are many other strategies that can be used in promoting the culture of assessment and continuous learning among educators and students. For example, schools and educational institutions can provide training and professional development opportunities for educators on the latest assessment methods and tools. These strategies may also encourage students to take ownership of their learning by involving them in the assessment process, such as through self-assessment or peer assessment.

Ultimately, developing new assessment models is essential for ensuring that Education 5.0 is successful in delivering high-quality, student-centered learning experiences. By developing innovative assessment methods, incorporating ongoing assessment into the learning process, and promoting a culture of assessment and continuous learning, we can help in ensuring that the students are prepared for the challenges and opportunities of the 21st century [64,166–168].

In conclusion, Education 5.0 represents a significant shift in educational models, which emphasizes the importance of personalized, collaborative, experiential, and globalized learning experiences. To fully realize the tremendous potential Education 5.0 has, we must address the challenges of the digital divide, educator training, and assessment models, and invest in the necessary infrastructure and resources to create a more equitable and effective education system.

13.5 SUMMARY AND CONCLUSIONS

The field of education has undergone significant changes over the past many years, with new approaches and paradigms emerging in response to the ever changing needs of learners and the society. In this article, we have explored several key trends and concepts in education including value-based education, research-based education, project-based education, experiential learning, student aspirations, flexibilities, Industry 5.0, curriculum design, teaching-learning-evaluation processes, outcome-based education, and the most important paradigm shift from teacher-centric learning to student-centric learning, and Education 5.0.

Value-based education emphasizes the importance of imparting values and ethics to learners, in addition to academic knowledge. Research-based learning highlights the importance of development research skills and critical thinking among the students as a part of their learning process. Project-based education involves learners in real-world projects, which help them to apply theoretical knowledge in practical situations. Experiential learning is basically a hands-on approach to learning, which provides learners with direct experience and exposure to real-world problems and situations.

Student aspirations are the individual goals and ambitions of learners, which should be considered when designing educational programs. Flexibilities in education refer to the ability to customize educational programs in such a way that they meet the individual needs and interests of learners. Industry 5.0 refers to the integration of emerging technologies and Industry 4.0 principles in the education system, and to prepare learners with the required skill set for the jobs of the future.

Curriculum design is an important aspect dealing with the process of designing educational programs, taking into account learner needs, societal needs, and emerging trends and technologies. Teaching-learning-evaluation processes refer to the methods and strategies used by educators to impart knowledge and evaluate learning outcomes. Outcome-based education focuses on clearly defined learning outcomes and assesses learners' abilities to achieve those outcomes.

The paradigm shift from teacher-centric learning to student-centric learning is an approach to education that focuses on the needs and interests of the learner, providing them with personalized and relevant learning experiences.

Education 5.0 is a future-focused educational paradigm that emphasizes the integration of emerging technologies, Industry 5.0 principles, and learner-centered approaches to education.

Education 5.0 is an innovative approach to education that focuses on holistic development of students and prepares them with skill set required to face the challenges of the 21st century. It emphasizes the importance of value-based education, research-based education, project-based education, and experiential learning. It also emphasizes the need for personalization, collaboration, and the integration of industry 5.0 into the education system.

Personalization is a crucial component of Education 5.0, allowing students to have customized learning paths, adaptive learning, flexible scheduling, and personalized assessments. Collaboration emphasizes the importance of peer-to-peer learning, group projects, community engagement, and virtual collaboration.

Technology plays a vital role in Education 5.0, with online learning platforms, virtual and augmented reality, artificial intelligence, cloud computing, and social media being some of the critical components.

Experiential learning focuses on internships, apprenticeships, service learning, project-based learning, study abroad programs, and experiential simulations, allowing students to gain hands-on experience and apply theoretical knowledge in real-world situations.

Globalization is another essential component of Education 5.0, with multicultural education, language learning, international study programs, global partnerships, and online global learning communities being critical components.

In terms of challenges and solutions, digital divide, training educators, and developing new assessment models are some of the most significant challenges that needed to be addressed. To overcome these challenges, it is essential to focus on providing equal access to technology, providing professional development for educators, and developing new assessment models that can measure the holistic development of students.

Education 5.0 is a transformational approach to education, which emphasizes the importance of holistic approach to student development, personalization, collaboration, technology, experiential learning, and globalization. While there are significant challenges to implementing this approach, with the right strategies and solutions, we can create an education system that prepares students for success in the 21st century.

In conclusion, the field of education is continuously evolving, with new approaches and paradigms emerging in response to the fast changing needs of learners and society. Value-based education, research-based learning, project-based learning, experiential learning, student aspirations, flexibilities, Industry 5.0, curriculum design, teaching-learning-evaluation processes, outcome-based education, paradigm shift from teacher-centric learning to student-centric learning and Education 5.0 are all important concepts that are shaping the future of education. By embracing these trends and concepts, educators can help prepare learners for success in the rapidly changing world.

REFERENCES

[1] UNESCO. (2019). *Education 5.0 and Artificial Intelligence for sustainable development*. Paris: UNESCO.
[2] UNESCO. (2019). *Futures of education: Learning to become*. UNESCO.
[3] Narayanan, S. (2016). *Value education in schools and colleges*. New Delhi: PHI Learning.
[4] Schwartz, S. H. (2017). An overview of the Schwartz theory of basic values. *Online Readings in Psychology and Culture*, 2(1), 11.
[5] De George, R. T. (2021). *Business ethics*. Routledge.
[6] Rokeach, M. (2017). *The nature of human values*. Routledge.
[7] Cortina, L. M., et al. (2020). *Universal human values and ethics: A cross-cultural approach*. Springer.

[8] Feinberg, J. (2019). *Moral philosophy: Theories and issues*. Cengage Learning.
[9] Gaur, R. R., Asthana, R., and Bagaria (2022). *Teacher's manual: A foundation course In human values and professional ethics (set of 2 books)*, 2nd Editions, Excel Books, New Delhi, India.
[10] Gaur, R. R., Asthana, R., and Bagaria (2010). *A foundation course in human values and professional ethics*, Excel Books, New Delhi, India.
[11] Kumar, V. (2019). *Values and ethics for organizations: Theory and practice*. SAGE Publications India.
[12] Healey, M., & Jenkins, A. (2009). *Developing undergraduate research and inquiry*. York, UK: Higher Education Academy.
[13] Thomas, J. W. (2000). *A review of research on project-based learning*. San Rafael, CA: Autodesk Foundation.
[14] Kolb, D. A. (2014). *Experiential learning: Experience as the source of learning and development*. Upper Saddle River, NJ: Pearson Education.
[15] Kolb, D. A. (2014). *Experiential learning: Experience as the source of learning and development*. FT press.
[16] Brown, M., & Jones, R. (2018). *Student aspirations and the transition to university: Widening participation and marginalisation*. New York: Routledge.
[17] Cleary, T. J., Callan, G. L., & Zimmerman, B. J. (2020). Academic aspirations and self-regulation. In *Self-regulation and motivation in learning environments* (pp. 59–78). Springer, Cham.
[18] Pate, J., & Short, P. M. (2019). Academic aspirations and achievement of first-generation college students: Implications for college counselors. *Journal of College Counseling*, 22(2), 165–176.
[19] Fester, M., & Hassenkamp, J. (2021). Career aspirations and professional identity development of university students in STEM disciplines. *Journal of Career Development*, 48(1), 48–63.
[20] Garg, P., & Malhotra, R. (2020). Career aspirations, social support, and career barriers among female students: A study of an Indian university. *Journal of Career Development*, 47(2), 160–174.
[21] Beyers, W., & Goossens, L. (2018). The reciprocal relationship between personal identity and personal aspirations: A longitudinal study in adolescence. *Journal of Research on Adolescence*, 28(1), 220–235.
[22] Côté, J. E., & Levine, C. G. (2017). *Identity formation, agency, and culture: A social psychological synthesis*. Psychology Press.
[23] Csikszentmihalyi, M. (2014). *Creativity: The psychology of discovery and invention*. Harper Perennial Modern Classics.
[24] Kaufman, J. C., & Sternberg, R. J. (Eds.). (2019). *The Cambridge handbook of creativity* (2nd ed.). Cambridge University Press.
[25] Crocetti, E., Klimstra, T. A., Hale, W. W., Koot, H. M., Meeus, W., & Branje, S. (2016). Friend similarity in group decision making: The roles of social identity and norms of cooperation. *Group Processes & Intergroup Relations*, 19(6), 730–749.
[26] Deci, E. L., & Ryan, R. M. (2018). Basic psychological needs theory, self-determination theory, and the importance of context. In H. T. Reis & S. K. Sprecher (Eds.), *Encyclopedia of human relationships* (2nd ed., pp. 167–171). Routledge.

[27] Shinnar, R. S., et al. (2019). Entrepreneurial aspirations and intentions: A large-scale study of youth across 20 countries. *Journal of Business Venturing Insights*, 12, e00132.
[28] Liao, Y., & Liu, Y. (2019). Entrepreneurial aspirations and entrepreneurial activities: The role of family background. *Journal of Entrepreneurship Education*, 22(1), 1–10.
[29] Fayolle, A., & Gailly, B. (2015). The impact of entrepreneurship education on entrepreneurial attitudes and intention: Hysteresis and persistence. *Journal of Small Business Management*, 53(1), 75–93.
[30] Astuti, P., & Triyono, D. (2019). The influence of entrepreneurial orientation on entrepreneurial aspirations and perceived behavioral control. *Journal of Entrepreneurship Education*, 22(1), 1–9.
[31] Dhiman, S., & Bhardwaj, A. (2021). Investigating the role of entrepreneurial aspirations in developing entrepreneurial intention: A study of Indian business students. *Journal of Entrepreneurship Education*, 24(2), 1–14.
[32] Cookson Jr, P. W. (2018). *Flexibility and public education*. New York: Teachers College Press.
[33] Baik, C., & Greig, J. (2009). Improving assessment practice through cross-institutional collaboration: From concept to reality. *Assessment & Evaluation in Higher Education*, 34(3), 303–313.
[34] Garrison, D. R., & Kanuka, H. (2004). Blended learning: Uncovering its transformative potential in higher education. *Internet and Higher Education*, 7(2), 95–105.
[35] Abele, E., Wortmann, F., Giger, Y., Meisen, T., Morisse, M., Urmetzer, F., & Neely, A. (2018). The future of work and education in Industry 5.0. *Journal of Industrial Information Integration*, 10, 1–8.
[36] Lee, J. Y., & Park, J. H. (2019). Industry 5.0: An overview of key technologies and their potential effects on businesses and society. *Technological Forecasting and Social Change*, 141, 341–351.
[37] Prajogo, D. I., & Oke, A. (2018). Industry 4.0 and manufacturing flexibility: An overview. *Emerald Insight*, 38(7), 1206–1224.
[38] Shin, H., & Kang, J. (2020). Industry 5.0 and the future of work: Challenges and opportunities for human resource development. *Sustainability*, 12(6), 2486.
[39] World Economic Forum. (2019). *The Future of Jobs Report 2018*. Geneva: World Economic Forum.
[40] Ghobakhloo, M., et al. (2021). Industry 5.0: State-of-the-art and future directions for sustainable development. *Journal of Cleaner Production*, 283, 124654.
[41] Brynjolfsson, E., & McAfee, A. (2014). *The second machine age: Work, progress, and prosperity in a time of brilliant technologies*. WW Norton & Company.
[42] Davenport, T. H., & Kirby, J. (2015). Beyond automation: Strategies for remaining gainfully employed in an era of very smart machines. *Harvard Business Review*, 93(6), 58–65.
[43] Dasgupta, S., Hettiarachchi, H., & Huq, M. (2017). Achieving sustainable development goals through sustainable manufacturing: A review and a conceptual framework. *Journal of Cleaner Production*, 162, 1209–1222.

[44] Figueiredo, J., & Ferreira, V. (2019). Sustainable manufacturing in the Industry 4.0 era: An overview of recent developments and future perspectives. *Journal of Cleaner Production*, 240, 118204.
[45] Lemon, K. N., & Verhoef, P. C. (2016). Understanding customer experience throughout the customer journey. *Journal of Marketing*, 80(6), 69–96.
[46] Gupta, S., Lehmann, D. R., & Stuart, J. A. (2004). Valuing customers. *Journal of Marketing Research*, 41(1), 7–18.
[47] Kotler, P. (2017). *Marketing 4.0: Moving from traditional to digital*. John Wiley & Sons.
[48] Batra, A., Singh, H., Singh, A., & Narkhede, B. E. (2021). *The Future of Smart Factories: Industry 4.0. In Handbook of Industry 4.0 and Smart Manufacturing* (pp. 1–22). Springer.
[49] Akram, R. N., & Kulathuramaiyer, N. R. (2019). Towards Industry 4.0: A comprehensive review on smart factories. *Journal of Industrial Information Integration*, 15, 100–119.
[50] Fink, L. D. (2016). *Creating significant learning experiences: An integrated approach to designing college courses*. San Francisco, CA: Jossey-Bass.
[51] Arredondo, M., & Gallardo, K. (2018). *Curriculum needs assessment: A primer for K-12 educators*. Routledge.
[52] Hiebert, J. (2018). A needs assessment for mathematics curriculum development: Considering alignment with implementation. *Educational Researcher*, 47(5), 319–328.
[53] Lattuca, L. R., & Stark, J. S. (2011). *Shaping the college curriculum: Academic plans in context*. John Wiley & Sons.
[54] Fink, L. D. (2013). *Creating significant learning experiences: An integrated approach to designing college courses*. John Wiley & Sons.
[55] Posner, G. J. (2004). *Analyzing the curriculum*. McGraw-Hill Education.
[56] Tyler, R. W. (1949). *Basic principles of curriculum and instruction*. University of Chicago Press.
[57] Wiggins, G. P., & McTighe, J. (2005). Understanding by design. Association for Supervision and Curriculum Development.
[58] Brookfield, S. D. (2015). *The skillful teacher: On technique, trust, and responsiveness in the classroom*. John Wiley & Sons.
[59] Daniels, H. (2018). *Designing tasks in secondary education: Enhancing subject understanding and student engagement*. Routledge.
[60] Gagne, R. M., Wager, W. W., Golas, K. C., & Keller, J. M. (2004). *Principles of instructional design*. Cengage Learning.
[61] McTighe, J., & O'Connor, K. (2005). Seven practices for effective learning. *Educational Leadership*, 63(3), 10–17.
[62] Stiggins, R. J. (2015). *Assessment for learning redefined: Using assessment to promote learning and achievement*. ASCD.
[63] Wiggins, G. P. (2012). Seven keys to effective feedback. *Educational Leadership*, 70(1), 10–16.
[64] Black, P., & Wiliam, D. (2009). Developing the theory of formative assessment. Educational Assessment. *Evaluation and Accountability*, 21(1), 5–31.
[65] Brookhart, S. M. (2013). *How to assess higher-order thinking skills in your classroom*. ASCD.

[66] Cohen, L., Manion, L., & Morrison, K. (2018). *Research methods in education*. Routledge.
[67] Best, J. W., & Kahn, J. V. (2016). *Research in education*. Pearson Education India.
[68] Adams, M. J. (2010). Benchmarking: The search for industry best practices that lead to superior performance. *Journal of Business Strategy*, 31(3), 28–38.
[69] Boyatzis, R. E. (2013). Competencies in the 21st century. *Journal of Management Development*, 32(1), 5–12.
[70] Kellaghan, T., & Greaney, V. (2016). *Using assessment to improve the quality of education*. Springer.
[71] Liker, J. K., & Meier, D. (2018). *The Toyota way: 14 management principles from the world's greatest manufacturer*. McGraw Hill Professional.
[72] McAfee, A., & Brynjolfsson, E. (2017). *Machine, platform, crowd: Harnessing our digital future*. WW Norton & Company.
[73] Raman, A. (2017). *The third pillar: How markets and the state leave the community behind*. Penguin.
[74] Marzano, R. J. (2007). *The art and science of teaching: A comprehensive framework for effective instruction*. ASCD.
[75] McTighe, J., & Wiggins, G. (2012). *Understanding by design framework*. ASCD.
[76] Shulman, L. S. (1987). Knowledge and teaching: Foundations of the new reform. *Harvard Educational Review*, 57(1), 1–22.
[77] Hattie, J. (2012). *Visible learning for teachers: Maximizing impact on learning*. Routledge.
[78] Schunk, D. H. (2012). *Learning theories: An educational perspective*. Pearson.
[79] Vygotsky, L. S. (1978). *Mind in society: The development of higher psychological processes*. Harvard University Press.
[80] Bransford, J. D., Brown, A. L., & Cocking, R. R. (2000). *How people learn: Brain, mind, experience, and school*. National Academies Press.
[81] Popham, W. J. (2009). *Assessment for educational leaders*. Pearson.
[82] Scriven, M. (1967). The methodology of evaluation. *American Educational Research Journal*, 4(4), 663–687.
[83] Stiggins, R. J. (2002). Assessment crisis: The absence of assessment for learning. *Phi Delta Kappan*, 83(10), 758–765.
[84] Guskey, T. R. (2003). How classroom assessments improve learning. *Educational Leadership*, 60(5), 6–11.
[85] Spady, W. G. (1994). *Outcome-based education: Critical issues and answers*. Arlington, VA: American Association of School Administrators.
[86] Goodlad, J. I. (1990). *Teachers for our nation's schools*. San Francisco, CA: Jossey-Bass.
[87] Blackwell, R. D. (1978). Education 1.0: A simple model of education in the 19th century. *Curriculum Inquiry*, 8(3), 255–269.
[88] Downes, S. (2005). E-learning 2.0. *eLearn magazine*, 2005(10), 1–4.
[89] Siemens, G. (2013). Massive open online courses: Innovation in education? In R. McGreal, W. Kinuthia, & S. Marshall (Eds.), *Open educational resources: Innovation, research and practice* (pp. 5–16). Commonwealth of Learning and Athabasca University.

[90] Siemens, G. (2014). Connectivism: A learning theory for the digital age. In K. Walsh, S. Osborne, & M. Williams (Eds.), *The SAGE handbook of e-learning research* (pp. 35–48). Sage Publications.
[91] World Economic Forum. (2016). The future of jobs: Employment, skills and workforce strategy for the fourth industrial revolution.
[92] Gupta, A., & Jangra, A. (2021). Education 4.0: The next wave of technology-enabled Eeducation. In *Handbook of Research on Innovative Pedagogies and Models for Educational Success* (pp. 1–18). IGI Global.
[93] Raza, S. A., & Nawaz, M. (2021). Industry 5.0 and education 5.0: Opportunities and challenges for future education systems. *International Journal of Emerging Technologies in Learning*, 16(3), 49–58.
[94] Wang, F., & Wang, J. (2020). Education 5.0 and the development of talent in the new era. *International Journal of Higher Education*, 9(2), 132–140.
[95] Alshahrani, A., & Alharbi, M. (2021). Education 5.0: An overview and its implications on higher education. *Journal of Education and Learning*, 10(1), 25–33.
[96] Hartshorne, R., & Cyboran, V. (2020). Education 5.0: Exploring the impact of artificial intelligence, machine learning, and robotics on education. *Journal of International Education Research*, 16(3), 57–66.
[97] Kolić-Vehovec, S., & Miljković, D. (2021). Education 5.0: Possibilities and limitations of smart technologies in education. *International Journal of Emerging Technologies in Learning*, 16(7), 114–125.
[98] Prinsloo, P., & Slade, S. (2017). An elephant in the learning analytics room: the obligation to act. *British Journal of Educational Technology*, 48(6), 1401–1415.
[99] Bicen, H., & Koc, M. (2018). The personalized education and e-learning system based on the big five personality traits. *Education and Information Technologies*, 23(1), 499–515.
[100] Gómez-Sánchez, E., Huertas-Valdivia, I., & Rubio-Romero, J. C. (2020). Personalized Learning and Big Data Analysis: A Proposal for Its Implementation in Education. In *Data Analysis in Education* (pp. 49–68). Springer, Cham.
[101] VanLehn, K. (2011). The relative effectiveness of human tutoring, intelligent tutoring systems, and other tutoring systems. *Educational Psychologist*, 46(4), 197–221. 10.1080/00461520.2011.611369.
[102] Kopecky, K. J., & Johnson, T. E. (2012). Designing personalized e-learning with learner-generated contexts. *Journal of Educational Technology & Society*, 15(3), 236–248.
[103] Dhawan, S. (2020). Online learning: A panacea in the time of COVID-19 crisis. *Journal of Educational Technology Systems*, 49(1), 5–22.
[104] Fitzpatrick, C., & Jones, C. (2019). *Flexible pedagogies: Technology-enhanced learning*. Association for Learning Technology.
[105] Anshari, M., Alas, Y., & Guan, L. (2018). Smart campus: Integrating technology enhanced active learning in higher education. In *Handbook of Research on Mobile Technology, Constructivism, and Meaningful Learning* (pp. 212–227). IGI Global.
[106] Mulder, R. H., & Sloep, P. B. (2017). Personalized assessment as a leverage point for effective learning: current developments and future directions.

Educational Technology Research and Development, 65(3), 555–571. 10.1 007/s11423-017-9511-8

[107] Karabulut-Ilgu, A., & Jaramillo Cherrez, N. (2020). Personalized assessment and feedback in online learning: A scoping review. *Online Learning*, 24(4), 1–26. 10.24059/olj.v24i4.2173.

[108] Hämäläinen, R., & Vesisenaho, M. (2017). Peer learning in higher education: An integrative literature review. *Higher Education*, 74(2), 221–240.

[109] Slavin, R. E. (1995). *Cooperative learning: Theory, research, and practice* (2nd ed.). Allyn & Bacon.

[110] Cox, M. D., & Brestan-Knight, E. (2020). Active and collaborative learning: Engaging students in the classroom. *Psychology Learning and Teaching*, 19(2), 211–227.

[111] Ehrlich, T. (2021). Community-engaged teaching in the time of COVID-19. *Journal of Community Engagement and Higher Education*, 13(1), 1–8.

[112] Bower, M. (2019). *The next generation of distance education: Unbundling and education technology*. Cham, Switzerland: Springer International Publishing.

[113] Bower, M., Dalgarno, B., Kennedy, G. E., Lee, M. J., & Kenney, J. (2015). Design and implementation factors in blended synchronous learning environments: Outcomes from a cross-case analysis. *Computers & Education*, 86, 1–17. doi: 10.1016/j.compedu.2015.02.013.

[114] Liyanagunawardena, T. R., Adams, A. A., & Williams, S. A. (2013). MOOCs: A systematic study of the published literature 2008-2012. *The International Review of Research in Open and Distributed Learning*, 14(3), 202–227.

[115] Al Lily, A. E., Ismail, A. F., & Abunasser, F. M. (2013). Massive open online courses (MOOCs): A review. *Journal of Educational Technology & Society*, 16(2), 58–67.

[116] Akçayır, M., & Akçayır, G. (2017). Advantages and challenges associated with augmented reality for education: A systematic review of the literature. *Educational Research Review*, 20, 1–11.

[117] Chen, Y. C., & Huang, Y. M. (2018). Enhancing learning effectiveness through the integration of augmented reality with self-regulated learning strategies. *Journal of Educational Technology & Society*, 21(2), 78–89.

[118] Hew, K. F., & Cheung, W. S. (2016). Use of Three-Dimensional (3-D) Immersive Virtual Worlds in K-12 and Higher Education Settings: A Review of the Research. *British Journal of Educational Technology*, 47(1), 1–19.

[119] Knight, J. (2016). The role of technology in the future of higher education. *Educause Review*, 51(1), 32–42.

[120] Li, J., Li, X., Li, D., & Li, Y. (2019). Integration of cloud computing into higher education: a systematic review. *Education and Information Technologies*, 24(4), 2293–2312.

[121] Huang, C. M., & Liang, T. H. (2019). A cloud-based mobile learning system for improving student learning achievement in higher education. *Journal of Educational Technology & Society*, 22(2), 12–24.

[122] Manca, S., & Ranieri, M. (2016). Is Facebook still a suitable technology-enhanced learning environment? An updated critical review of the literature from 2012 to 2015. *Journal of Computer Assisted Learning*, 32(6), 503–528. 10.1111/jcal.12154.

[123] Junco, R. (2012). The relationship between frequency of Facebook use, participation in Facebook activities, and student engagement. *Computers & Education*, 58(1), 162–171. 10.1016/j.compedu.2011.08.004.
[124] Brown, L. E. (2015). *Experiential learning: A best practice handbook for educators and trainers*. Kogan Page Publishers.
[125] Banks, J. A. (2006). *Cultural diversity and education: Foundations, curriculum, and teaching* (5th ed.). Pearson Education.
[126] Ramos-Sánchez, L., & Nichols, L. (2013). Multicultural education: A renewed paradigm of transformation and call to action. *Journal of Educational Research and Practice*, 3(2), 13–26.
[127] Sleeter, C. E. (2011). Confronting the marginalization of culturally responsive pedagogy. *Urban Education*, 46(4), 790–796.
[128] Gay, G. (2010). *Culturally responsive teaching: Theory, research, and practice* (2nd ed.). Teachers College Press.
[129] Baker, C. (2011). *Foundations of bilingual education and bilingualism* (5th ed.). Multilingual Matters.
[130] Lightbown, P. M., & Spada, N. (2013). *How languages are learned* (4th ed.). Oxford University Press.
[131] Zhang, J., & Barber, B. (2017). Technology-enhanced language learning for intercultural communication: A review. *Educational Technology Research and Development*, 65(4), 893–918.
[132] Zhao, H., et al. (2018). Entrepreneurial aspirations and firm performance in emerging markets. *Journal of Business Venturing*, 33(2), 117–131.
[133] Cook, V. (2016). *Second language learning and language teaching* (5th ed.). Routledge.
[134] García, O., & Li, W. (2014). *Translanguaging: Language, bilingualism and education*. Palgrave Macmillan.
[135] Mäkelä, L., & Ruohotie-Lyhty, M. (2016). Assessing the impact of internationalisation on teaching and learning in higher education. *Journal of Studies in International Education*, 20(5), 415–431.
[136] Wilkins, S., & Balakrishnan, M. S. (2013). *Internationalisation of higher education: Changing paradigms and approaches*. SAGE Publications.
[137] Dwyer, M. M. (2019). The impact of study abroad on student success. *Frontiers: The Interdisciplinary Journal of Study Abroad*, 31(1), 1–14.
[138] Goh, M., & Woods, P. C. (2012). *Enhancing the internationalisation of higher education: Globalisation, internationalisation and cross-border education*. Routledge.
[139] Leask, B., & Bridge, C. (2013). *Internationalising the curriculum in the disciplines: Imaginings, understandings and practices*. Routledge.
[140] Van de Ven, A. H., & Johnson, P. E. (2006). Knowledge for theory and practice. *Academy of Management Review*, 31(4), 802–821.
[141] Deardorff, D. K. (2011). Assessing intercultural competence in higher education: Existing research and future directions. *International Journal of Intercultural Relations*, 35(3), 377–389.
[142] Knight, J. (2014). Internationalization remodeled: Definition, approaches, and rationales. *Journal of Studies in International Education*, 18(3), 204–219.
[143] Levy, M. (2007). Culture, culture learning and new technologies: Towards a pedagogical framework. *Language Learning & Technology*, 11(2), 104–127.

[144] Warschauer, M., & Grimes, D. (2008). Audience, authorship, and artifact: The emergent semiotics of Web 2.0. *Annual Review of Applied Linguistics*, 28, 1–18.
[145] Kanuka, H., & Anderson, T. (2010). Online social interchange, discord, and knowledge construction. *Journal of Distance Education*, 24(1), 21–34.
[146] Kirschner, P. A., & Davis, N. E. (2003). Pedagogical agents: Overcoming the lack of human presence in e-learning environments. *Educational Technology, Research and Development*, 51(3), 47–56.
[147] Lifelong Learning Council Queensland. (2018). What is lifelong learning? Retrieved from https://www.llcq.org.au/what-is-lifelong-learning/
[148] Peters, M. A., & Besley, T. (Eds.). (2020). *Handbook of education policy studies*. Springer.
[149] Rogers, E. M. (2003). *Diffusion of innovations*. Simon and Schuster.
[150] Bélanger, P., & Lin, C. (2012). A measure of lifelong learning readiness. *International Journal of Lifelong Education*, 31(4), 393–410.
[151] UNESCO. (2013). *The role of lifelong learning in the 21st century*. Retrieved from https://unesdoc.unesco.org/ark:/48223/pf0000223929
[152] UNESCO. (2019). *Education 2030: Incheon declaration and framework for action for the implementation of sustainable development goal 4*. Paris: UNESCO.
[153] Yorke, M. (2006). Lifelong learning in higher education: A critical appraisal. *Learning and Teaching in Higher Education*, 1(1), 3–16.
[154] OECD. (2019). *OECD digital education outlook 2019*. OECD Publishing.
[155] Warschauer, M. (2003). *Technology and social inclusion: Rethinking the digital divide*. MIT Press.
[156] Wertsch, J. V. (2007). Mediation. In *The Cambridge handbook of sociocultural psychology* (pp. 207–220). Cambridge University Press.
[157] Bimber, B. (2000). Measuring the gender gap on the internet. *Social Science Quarterly*, 81(3), 868–876.
[158] Dutta, S., & Mia, I. (2017). *Global information technology report 2016*. In World Economic Forum.
[159] Kirschner, P. A., & De Bruyckere, P. (2017). The myths of the digital native and the multitasker. *Teaching and Teacher Education*, 67, 135–142.
[160] Timperley, H., Wilson, A., Barrar, H., & Fung, I. (2007). *Teacher professional learning and development: Best evidence synthesis iteration [BES]*. Ministry of Education.
[161] Borko, H., & Livingston, C. (1989). Cognition and improvisation: Differences in mathematics instruction by expert and novice teachers. *American Educational Research Journal*, 26(4), 473–498.
[162] Darling-Hammond, L. (2017). Teacher education around the world: What can we learn from international practice?. *European Journal of Teacher Education*, 40(3), 291–309.
[163] Fullan, M. (2001). *The new meaning of educational change*. Routledge.
[164] Guskey, T. R. (2002). Professional development and teacher change. *Teachers and Teaching*, 8(3), 381–391.
[165] Hargreaves, A., & Fullan, M. (2012). *Professional capital: Transforming teaching in every school*. Teachers College Press.

[166] Leighton, J. P., & Gierl, M. J. (2007). The learning and assessment of knowledge and skills in educational contexts. *Review of Research in Education*, 31(1), 275–313.
[167] National Research Council. (2011). *Assessment in the service of learning: Principles and practices for effective assessment*. National Academies Press.
[168] Ferguson, R., & Buckingham Shum, S. (2012). Learning analytics to identify exploratory dialogue within synchronous text chat. *Distance Education*, 33(1), 105–121.